Palgrave Studies in the History of Science and Technology

Designed to bridge the gap between the history of science and the history of technology, this series publishes the best new work by promising and accomplished authors in both areas. In particular, it offers historical perspectives on issues of current and ongoing concern, provides international and global perspectives on scientific issues, and encourages productive communication between historians and practicing scientists.

More information about this series at
http://www.palgrave.com/gp/series/14581

William A.T. Logan

A Technological History of Cold-War India, 1947–1969

Autarky and Foreign Aid

palgrave
macmillan

William A.T. Logan
Department of History
Pacific Union College
Angwin, CA, USA

ISSN 2730-972X ISSN 2730-9738 (electronic)
Palgrave Studies in the History of Science and Technology
ISBN 978-3-030-78766-0 ISBN 978-3-030-78767-7 (eBook)
https://doi.org/10.1007/978-3-030-78767-7

Cover image: US National Archives

This Palgrave Macmillan imprint is published by the registered company Springer Nature Switzerland AG
The registered company address is: Gewerbestrasse 11, 6330 Cham, Switzerland

To my parents

Acknowledgments

Like many academic books, this one has been in the making for literally a decade. This has been a long journey, and along the way, many people have advised, encouraged, and otherwise supported me. Here, I give thanks.

When I started this project at Auburn University, my advisor Alan Meyer supported and encouraged me at every step. I miss our regular meetings over tea at Mama Mocha's in Auburn. During the last phases of writing at my current academic home of Pacific Union College, my department chair Howard Munson was a dependable source of humor and sanity as California burned and a pandemic raged. Other fellow historians who provided feedback or support on this project at some stage include David Carter, Martin Collins, Gail Cooper, David C. Engerman, James Hansen, Arne Kaijser, Ralph Kingston, David Lucsko, Kenneth Noe, Mark Sheftall, and William Trimble. Thank you all! At Palgrave Macmillan, I am thankful to Roger Launius, Meagan Simpson, and Shreenidhi Natarajan for their help and support.

I remain forever indebted to the hardworking staff of the Ralph Brown Draughon Library at Auburn University, especially the interlibrary loan department. Without them, I never would have gotten my hands on microfilms of half-century-old newspapers or PDFs of obscure engineering journals. Later home-base libraries, at Walla Walla University and Pacific Union College, also provided me invaluable support during later stages of research.

I am also thankful to these other libraries and archives in the United States that opened their doors to me: National Archives and Records Administration–College Park, Library of Congress (especially the Asian Division), Air Force Historical Research Agency at Maxwell Air Force Base, the Water Resources Collection and Archives at the University of California–Riverside, the Butler Library at Columbia University, and the Doe Library at the University of California–Berkeley.

Research in the United States was, of course, only half the story. I am thankful to the following institutions in India for assisting my research: National Archives of India, Nehru Memorial Museum and Library, National Library of India, Assam State Archives, Meghalaya State Central Library, Meghalaya Energy Corporation Ltd., and Northeast Council Library. I would especially like to thank Meeta Deka of Gauhati University and Jayanta Sharma of the Northeast Frontier Railway for providing me guidance and assistance during my research in Assam. *Bahut shukriya!*

My thanks to the following for providing grants that helped cover the cost of my research: the Society for Historians of American Foreign Relations, my own History Department at Auburn University, the Auburn University Graduate School, and Pacific Union College.

To the many people who offered me a place to stay during my research peregrinations, I give my thanks: Andrew and Jéssie Semotiuk in California, Ted Logan and Gem Stone-Logan in California, Sharon Ford in Maryland, Bethany and Joshua Ropa in New York, Jonathan Shafer in Washington, DC, Venky Mannu and Usha Davamony in Delhi, and Barnabas Sangma and Jemima Marak in Meghalaya.

Lastly, my thanks to Alex Keene for his humor and unflagging encouragement. To Andrew Jabola, for all the aerospace videos. To John Mohr, for stimulating discussions about VIS. To Ryan and Kayla Anderson, for being Ryan and Kayla Anderson. To my parents, Douglas and Suzan Logan, for everything: for giving me your engineering and history genes back in the eighties, and more recently, for always providing me a place to land in between jobs or journeys to India. And to Verónica Trinidad Gallegos Logan, who has been a faithful companion and dedicated advocate of my writing and research ever since we began our journey together. *Tu hai meri jaan-e-jaan, meri zindagi.*

Contents

1	**Introduction**	1
	Historiography	5
	Overview of the Book	9
	Arguments	10
	References	13
2	**The Industrial Revolution in India**	15
	Industry and the Raj	18
	Dreams of a New India	20
	The Five-Year Plans	25
	Independence and Autarky	29
	Import Substitution and Nehruvian India	31
	Waste and Adaptive Reuse in Indian Technology	37
	Conclusion	40
	References	40
3	**India and the Cold War**	43
	Nonalignment in India and Abroad	45
	India and the United States	50
	India and the Soviet Union	61
	Cultural Exchanges Between the Superpowers and India	66
	Conclusion	69
	References	71

4 Modernizing the Indian Air Force: Imported Fighters and the HF-24 Marut, 1947–1968 73
The Indian Air Force Before Independence 78
The Air Force Looks West 81
The Indigenous Fighter Dream 85
The Air Force and the Superpowers 89
India, Egypt, and Indigenous Jet Fighters 95
Conclusion 99
References 100

5 India's Air Defenses and the Sino-Indian War: Exercise Shiksha, November 1963 103
India and China Go to War 106
The West and India's Air Defenses 113
The Air Force Gets an Education 117
Friends and Rivals in Exercise Shiksha 123
From Palam Field to Kanjarkot Fort 127
Conclusion 129
References 132

6 Spanning Assam's Great River: The Brahmaputra Bridge and the Indian Railways, 1958–1962 135
The Indian Railways and Assam 138
Prehistory of the Brahmaputra Bridge 145
The Search for a Second Contractor 151
Building the Brahmaputra Bridge 157
Conclusion 163
References 166

7 A Dam in the Khasi Hills: The Umiam Hydroelectric Project, 1960–1965 169
Dams in Independent India 172
A Dam on Khasi Land 175
How the Dam Was Built 181
How the Dam Was Financed 187
Conclusion 193
References 197

8 Tarapur and the Atomic Age, 1959–1964 199
The Nuclear Age Comes to Independent India 202
The Genesis of Tarapur 206

The Race for the Tarapur Contract 210
Tarapur and the Cold-War World 215
Understanding Tarapur 220
References 221

9 Constructing Tarapur, 1964–1969 223
The Makers of Tarapur Atomic Power Station 226
Work at Tarapur 230
Life at Tarapur 237
Labor and the Indian State 245
The Tarapur Strike 249
Tarapur's Troubles 253
Conclusion 257
References 259

10 Conclusion 261
The Technological Heritage of Nehruvian India 265
Technology Transfer and Indigenization in Modern India 268
Nehruvian Development in Retrospect 270
References 274

Index 277

Acronyms and Abbreviations

AFHRA	Air Force Historical Research Administration (Maxwell AFB, AL, USA)
APED	Atomic Power Equipment Department (General Electric)
ASEB	Assam State Electricity Board
AT	*Assam Tribune*
BBJ	Braithwaite Burn & Jessop Construction Co. Ltd.
BWR	Boiling Water Reactor
CongRec	*Congressional Record*
CWPC	Central Water and Power Commission
DLF	Development Loan Fund
FRUS	*Foreign Relations of the United States*
GE	General Electric
GOI	Government of India
HCC	Hindustan Construction Co. Ltd.
HT	*Hindustan Times*
IAEA	International Atomic Energy Agency
IAF	Indian Air Force
IE	*Indian Express*
IJPRVD	*Indian Journal of Power and River Valley Development*
INTUC	Indian National Trade Union Congress
LSD	Lok Sabha Debates
NAI	National Archives of India (New Delhi)
NARA	National Archives and Records Administration (College Park, MD, USA)
NFR	Northeast Frontier Railway
NLI	National Library of India (Kolkata)

NMML	Nehru Memorial Museum and Library (New Delhi)
NYT	*New York Times*
RAAF	Royal Australian Air Force
RAF	Royal Air Force (United Kingdom)
RAPS	Rajasthan Atomic Power Station
RIAF	Royal Indian Air Force
TAPP	Tarapur Atomic Power Project
TAPS	Tarapur Atomic Power Station
TOI	*Times of India*
USAEC	United States Atomic Energy Commission
USAF	United States Air Force
USAID	United States Agency for International Development
USIS	United States Information Service
WRCA	Water Resources Collection and Archives (University of California—Riverside)

List of Figures

Fig. 1.1	Map of India showing locations of the case studies and other significant places mentioned in the following chapters. Note that while several Indian cities have undergone name changes in recent decades, this book uses the geographical names that were in use in the early independence period. Modern names of cities, where applicable, are given in parentheses on the map (Drawn by the author)	8
Fig. 3.1	Prime Minister Nehru with President Kennedy at the Indian leader's arrival in Washington on a November 1961 visit. In the back row are B.K. Nehru, Indian Ambassador to the United States (and the Prime Minister's cousin); US Secretary of State Dean Rusk; and US Vice President Lyndon B. Johnson (*Source* JFK Library, JFKWHP-1961-11-06-A)	53
Fig. 3.2	President Eisenhower on his visit to India in December 1959. Here he addresses a huge crowd at Ram Lila Maidan in Delhi (*Source* U.S. Embassy New Delhi Flickr account)	57
Fig. 3.3	US First Lady Jacqueline Kennedy visiting Nehru at Teen Murti Bhawan, his house in New Delhi, March 1962. Also present are Nehru's daughter Indira Gandhi and US Ambassador to India J.K. Galbraith and his wife Kitty (*Source* JFK Library, JFKWHP-1962-03-13-H)	60

Fig. 4.1 An ex-IAF B-24 liberator, on display at the Pima Air and Space Museum in Tucson, Arizona, in 2011 (Photo by the author) 82

Fig. 4.2 An IAF C-119G packet, modified with a booster engine on the top of the fuselage, on display at the IAF Museum in New Delhi, 2012 (Photo by the author) 84

Fig. 4.3 HF-24 Marut D-1698, a two-seater trainer version, on display at the HAL Museum in Bangalore in 2015 (Photo by the author) 87

Fig. 4.4 An IAF MiG-21, on display at the IAF Museum in New Delhi, 2012 (Photo by the author) 91

Fig. 5.1 The western portion of the China–India border, showing the contested area (Map drawn by author, based on map at http://www.lib.utexas.edu/maps/middle/east/and/asia/china/indiaw/border/88.jpg, accessed March 2, 2012) 108

Fig. 5.2 The eastern section of the China–India border, showing the contested area (Map drawn by author, based on map at http://www.lib.utexas.edu/maps/middle/east/and/asia/china/india/e/border/88.jpg, accessed March 2, 2012) 109

Fig. 5.3 A visibly haggard Nehru conferring with US Ambassador Galbraith at the time of the border war with China (*Source* JFK Library, JFKPOF-118a-004) 112

Fig. 5.4 American mobile radar sets in use during exercise Shiksha: AN/MPS-11 search (left) radar and AN/MPS-16 height-finder (right) (*Source* "Operation Shiksha," Moving Images Relating to Military Aviation Activities 1909–1984, RG 342, NARA) 117

Fig. 6.1 Maps of rail lines in northeast India. Before Partition (*top*), the rail lines to Assam ran through eastern Bengal. After Partition (*bottom*), when eastern Bengal became East Pakistan, new lines had to be constructed to connect Assam with the rest of India through the Siliguri Corridor (Drawn by the author) 141

Fig. 6.2 Map of the Lower Brahmaputra in the 1960s (Map drawn by the author, based on US Army Map Service maps at http://www.lib.utexas.edu/maps/ams/india/, accessed April 21, 2016) 143

Fig. 6.3 Diagram of a typical span of the Saraighat Bridge (Drawing by the author, based on diagrams and data in S. Ponnuswamy and V.C. Sharma, eds., *Bridging River Brahmaputra: Past Present and Future* [Guwahati: Northeast Frontier Railway, 1988], esp. p. 30) 159
Fig. 6.4 The Sankosh River Bridge on the Assam Rail Link in 2015. The spans are single Warren trusses (Photo by the author) 160
Fig. 6.5 Saraighat Bridge in 2015 (Photo by the author) 163
Fig. 7.1 Map of Umiam Lake and surroundings (Map drawn by the author, based on map in Assam State Electricity Board, "Umiam Hydro-Electric Project," commemorative pamphlet, n.d. [1966], Office of Director [Generation], Meghalaya Energy Corporation Ltd., Shillong) 180
Fig. 7.2 The dams of Umiam Hydroelectric Project (Drawing by the author, based on pictures in Assam State Electricity Board, "Umiam Hydro-Electric Project," commemorative pamphlet, n.d. [1966], Office of Director [Generation], Meghalaya Energy Corporation Ltd., Shillong) 183
Fig. 7.3 Aerial view of Umiam Lake in 2010 (Photo by the author) 192
Fig. 8.1 Map of the North Konkan Coast around Tarapur (Map drawn by the author, based on map in M.N. Chakravarti and M.R. Srinivasan, "Selection of Site for the Tarapur Atomic-Power Station," in International Atomic Energy Agency, *Siting of Reactors and Nuclear Research Centres* [Vienna: International Atomic Energy Agency, 1963], 457–65) 208
Fig. 9.1 The main gate of the Tarapur Atomic Power Project site, ca. 1969. The ventilation stack and reactor building are visible in the background (*Source* Photographs of Assistance Programs in Foreign Countries, RG 286-CP, NARA) 224
Fig. 9.2 One of two huge General Electric transformers installed on either side of the turbine-generator building at Tarapur Atomic Power Station. The shield on the cabinet door is a USAID symbol (*Source* Photographs from the Country Files, RG 286-CF, NARA) 229
Fig. 9.3 Indian technicians working inside one of the TAPS reactors (*Source* Photographs from the Country Files, RG 286-CF, NARA) 232
Fig. 9.4 View of the cooling water intake structure on the west side of the plant (*Source* Photographs of Assistance Programs in Foreign Countries, RG 286-CP, NARA) 234

Fig. 9.5 Maheshwar Dayal, project engineer, standing below the control rods of one of the reactors at Tarapur Atomic Power Station (*Source* Photographs from the Country Files, RG 286-CF, NARA) 237
Fig. 9.6 Don T. Spangenberg and G.J. Larsen of General Electric in the Tarapur control room (*Source* Photographs of Assistance Programs in Foreign Countries, RG 286-CP, NARA) 239

CHAPTER 1

Introduction

It was a sunny morning in October 1963 in the Shivalik foothills of northwestern India. A newly completed dam, Bhakra by name, stood in the sunlight, spanning the gorge of the Satluj River. At the top of the dam, a crowd of 50,000 people had gathered that morning, sitting on folding chairs set up on the 30-foot roadway on the dam's crest or perching, leaning, or standing on any other available vantage point on the hills on either side of the dam. The date was October 22, and the occasion was the dedication of the completed Bhakra Dam after fifteen long years of construction.[1]

The massive concrete dam—1700 feet long from one side of the gorge to the other and 560 feet tall from original ground level to crest—was one part of a major combined irrigation and hydroelectricity project. The project also included the much smaller Nangal Dam, eight miles downstream from Bhakra, which diverted some of the water of the Satluj into

[1] My account of the dedication of Bhakra Dam is based on the following contemporary accounts: "Bhakra Dam Dedicated to Nation by Nehru," *Times of India* (*TOI*), October 23, 1963; "Nehru Dedicates Bhakra Project to the Nation," *Hindustan Times* (*HT*), October 23, 1963; "Nehrūjī dwārā Bhākhṛā Bāndh rāshṭra ko samarpit: Viśāl jal-vidyut yojna sāmuhik prayatna kā pratīk," *Hindustān*, October 23, 1963; "Bhakra Dam Dedicated to the Nation," *India News*, October 25, 1963; "Highlights of the Week: The Bhakra Project," *Eastern Economist*, November 1, 1963, 891.

W. A. T. Logan, *A Technological History of Cold-War India, 1947–1969*, Palgrave Studies in the History of Science and Technology, https://doi.org/10.1007/978-3-030-78767-7_1

two concrete-lined irrigation channels to supply water to farms as far away as Rajasthan. For electric power generation, the project had a total of four powerhouses: two large ones at the foot of Bhakra Dam itself, and two smaller ones on one of the irrigation channels downstream from the dam. Except for the right-bank powerhouse at Bhakra, which was still being designed, the entire Bhakra-Nangal project was complete by October 1963.[2]

One of the speakers at the dedication of Bhakra Dam was Jawaharlal Nehru, Prime Minister of India and Chairman of the Planning Commission. A fervent advocate of modernization and development, Nehru had taken a keen interest in the Bhakra project throughout its construction, making trips to the site once every year or two. Always clad in his signature white cap, long jacket with a square collar and a fresh rose inserted in a buttonhole, and white leggings, the Prime Minister would inspect models of the project, peer into the Nangal Canal, and talk with engineers and technicians on his regular visits. On the day of the dedication, Nehru arrived by helicopter from Chandigarh. The Prime Minister took a full tour of Bhakra Dam, visiting the completed left-bank powerhouse and interior galleries of the concrete structure before riding an elevator up to the top of the dam.

In view of the crowd at the top of the dam, Nehru pushed a button to release water from Gobind Sagar, the reservoir behind the dam. The crowd burst into applause when water began to flow down the twin spillways on the face of the dam. Nehru addressed the crowd from a 20-foot-high stage erected on the western side of the dam's crest. The eloquent but diffusive Prime Minister declared the dam's completion an important national accomplishment, a preview of still greater achievements yet to come. He congratulated the people who had built the dam by their own physical and mental labor, and he paid homage to the 150 laborers killed in accidents during construction. He singled out for special commendation Harvey Slocum, a self-taught American engineer who had

2 "Dedication of Bhakra Dam," *Indian Recorder and Digest*, November 1963, 6. The figure of 560 feet for the height of the dam comes from an engineering drawing of "Bhakra Dam: Central Overflow Spillway Maximum Section" printed in Indian National Committee on Large Dams, *Design and Construction Features of Selected Dams in India* (New Delhi: Central Board of Irrigation and Power, 1979), 15. (The exact figures are 1700.25 ft [roadway elevation] – 1139.00 ft [original ground level] = 561.25 ft.) A more commonly printed figure, 740 ft / 226 m, is measured from the deepest foundations to the crest.

worked on New Deal dams in his home country before working for the Indian government on the Bhakra-Nangal project from 1951 until his death in 1961.[3] The Prime Minister also made reference to the military threat that India faced from China, and he encouraged his fellow Indians to "face this threat boldly and courageously."

As big as it was, the Bhakra-Nangal project was just one part of a decades-long modernization and development crusade launched by the Indian government after independence. Under the direction of the Planning Commission, the newly independent nation built new rail lines, power stations, irrigation works like the Nangal Canal, and factories for import substitution. The planners hoped that by industrializing, India would gain economic and technological autarky and thus be able to shake off the last vestiges of colonialism.

A key feature of this modernization drive was the adoption and—ideally—indigenization or Indianization of modern industrial technology from Europe, North America, or Japan. Nehruvian India's modernizing elites wanted their country to have the best that modern technology had to offer, but for that they would have to import some things: the machinery or technological artifacts themselves, the knowledge of how to use and how to make this technology, and sometimes even the capital to purchase it through foreign aid loans.

This was the case for the Bhakra-Nangal project. Although the project was built by the Indian government with Indian labor, the engineering team included some foreign specialists like Harvey Slocum, who brought their experience and expertise to the project. Foreign machinery also had a central role in the construction. The generators in the original left-bank powerhouse came from the Soviet Union. Transformers, switchgear, cranes, and earth-movers were also all imported. Even the Jeeps and Buicks that the project engineers drove to the jobsite had to be bought abroad.

When Bhakra Dam was new, it was a powerful symbol of modern India. As *Eastern Economist* said of the dam in November 1963: "It is a symbol of the nation's unity, its struggle against poverty and hunger and of the growing confidence to undertake mighty tasks." And industrializing India was truly a mighty task. Prime Minister Nehru's own rhetoric about the dam bordered on the religious. At the dedication ceremony, he said that

[3] "Harvey Slocum, Dam Builder, Dies," *New York Times* (*NYT*), November 12, 1961.

the dam was one of the "new temples of India." When the dam was still under construction, *Life* magazine quoted Nehru comparing the dam with India's most famous monument: "The Taj Mahal is for the dead," he is supposed to have remarked; "Bhakra is for the living."[4]

Yet while Bhakra Dam is a symbol, it is also much more than that. Bhakra is *technology*. It is concrete, steel, pipes, cables, generators, transformers, electromagnetism, fluid dynamics. Similarly, the Nehruvian development of India, while rooted in ideology and laden with symbolism, was a technological undertaking.

This book is a story of that undertaking. As the title indicates, it is a technological history, a history of the development projects and import-substitution initiatives of India in the first decades after independence. Most narratives of Nehruvian modernization focus on intellectuals like the influential economist P.C. Mahalanobis, the thinkers of development. This book focuses on the doers: the engineers, industrialists, government bureaucrats, diplomats, and common laborers—many anonymous in the historical record—who built Nehruvian India's planned development projects.[5]

The Cold War provides important context for understanding why Nehruvian modernization happened the way that it did. During this bitter struggle between the United States and the Soviet Union, nonaligned India strove to establish open relations with both superpowers. Starting in the mid-1950s, the superpowers began to wage the Economic Cold War, a contest over which country could develop the nonaligned nations first. As the biggest ex-colony and a major regional power, India was a key battleground of this struggle. Without the Economic Cold War,

[4] "Highlights of the Week: The Bhakra Project," *Eastern Economist*, November 1, 1963, 891; "Bhakra Dam Dedicated to the Nation," *India News*, October 25, 1963; "India Builds a High Dam in the Himalayas," *Life*, November 3, 1958, 45.

[5] For example, see "The Conquest of Nature," in Ramachandra Guha, *India After Gandhi: The History of the World's Largest Democracy* (New York: Ecco, 2007), 209–32. Another example of the intellectual-history trend of Indian development is Benjamin Zachariah, *Developing India: An Intellectual and Social History* (New Delhi: Oxford University Press, 2012). My approach in this book is to study "development-as-practiced," as David C. Engerman called for in his presidential address at the 2016 annual meeting of the Society for Historians of American Foreign Relations; see David C. Engerman, "Development Politics and the Cold War," *Diplomatic History* 41, no. 1 (January 2017): 1–19.

India would not have gotten as much attention or received as much development aid from the superpowers as it did.[6]

Nehruvian modernization was continually pulled in opposing directions, between a high ideal and gritty reality. The ideal was autarky, the idea that the Indian economy could be made, in the words of the Third Five-Year Plan, "self-reliant and self-generating." The reality was the need for foreign aid before India could develop. Paradoxically, in order to free itself from foreign domination, India accepted foreign aid in the form of loans and technical assistance.

The narrative of this book covers a 22-year period, from Indian independence in 1947 to 1969, when the scaled-back Fourth Five-Year Plan launched and one of the most ambitious development projects of the age, Tarapur Atomic Power Station, entered commercial operation. This was a period of optimism and idealism, a time when planners dreamed big and government departments and contractors built big. It was also a time of naïveté, because not even the most ambitious technocrats could engineer away economic constraints or the laws of nature. By the mid-sixties, Nehruvian modernization had foundered as the country redirected its resources to militarization and the specter of famine stalked the land.

Historiography

This book began as a response to a remarkable book by Daniel Headrick, *The Tentacles of Progress: Technology Transfer in the Age of Imperialism, 1850–1940*.[7] In this wide-ranging study of the interactions between European technology, European economies, and colonial Asian and African economies, Headrick argues that imperial governments retarded colonial development by focusing on infrastructure rather than education of the native population. From railroads to sanitation systems to quinine, Headrick tracks how technology supported colonial domination even while these technologies mostly failed to take root and be produced domestically.

[6] See David C. Engerman, *The Price of Aid: The Economic Cold War in India* (Cambridge, MA: Harvard University Press, 2018).

[7] Daniel Headrick, *The Tentacles of Progress: Technology Transfer in the Age of Imperialism, 1850–1940* (New York: Oxford University Press, 1988).

I first read *Tentacles of Progress* in a historiography of technology seminar in my first semester in graduate school at Auburn University. When I took the class, I had recently returned from my first sojourn in India, where I had spent a year as a volunteer schoolteacher in the Garo Hills of Meghalaya. Back in North America now, my mind was full of questions about why India was the way it was. Where I had lived in the Garo Hills, some farmers still tilled their rice fields with ox-driven plows, and many people lived in bamboo houses with dirt floors. And yet there was much modernity in evidence as well. Even in the rural Garo Hills, everybody seemed to have a mobile phone. The *Calcutta Telegraph*, which could be bought in the nearby town, carried American celebrity news.

During my time in the Garo Hills, I had noticed plenty of evidence that India produced some of the trappings of modernity domestically. At the local weekly market, almost all of the consumer goods for sale had MADE IN INDIA stamped on them. I rode to and from market every Wednesday morning in the school Sumo, an SUV manufactured in India by Tata Motors. India clearly had a large industrial sector, but how could that coexist in a country with bamboo huts and ox-driven plows?

Tentacles of Progress answered some of my questions but left me with others. The India of 1940, when the book's narrative ends, was hardly the India of 2010 that I knew. What had happened in the intervening seventy years? How did India go from being a colony with limited mechanized industry to a major industrial power? For those technologies that did take root and did become Indianized, how and why did that happen?

This book represents my attempt to answer these questions. While my book picks up where *Tentacles of Progress* leaves off, in the end it is not really a sequel to Headrick's work. *A Technological History of Cold-War India* is based on primary-source research and focuses exclusively on India in the early independence period. A wide-ranging work à la *Tentacles of Progress* that covers other former colonies is not really possible at this point, because there is not yet enough secondary literature to support such a study.[8] It is my hope that my book will someday serve as a building

[8] A shortage of secondary literature is a chronic condition for the historiography of postindependence India in general, and not just in the field of history of technology. As Ramachandra Guha argues, popular Indian historiography portrays independence as a climax and a culmination, after which nothing important could happen because the nation had already arrived at its goal. While Indian social scientists continued to study

block for a synthesis of technology and indigenization in the nonaligned world.[9]

Why study technology in India in the early independence period? Apart from filling a chronological gap in the historiography after Headrick's book, what does such an account add? There was indeed much continuity in terms of technology and development between the colonial and early independence periods. India was led by modernizing elites with European educations and scientific outlooks. Yet however much the modernizing elites of India and other ex-colonies had imbibed western ways of thinking, they were not Europeans; they were Asians or Africans, nationalists who wanted to develop their countries for what they thought would be the good of the nation, not for the benefit of a colonial power. Autarky was a key element of postindependence modernization that had been missing in economic programs during the colonial period. Indigenization and import-substitution programs grew out of the desire for autarky, and they reflected a new conception of technology that would not have been possible during the colonial period.

If Daniel Headrick suggested the framing for this book, it was David Edgerton who provided its theoretical background. In his 2007 book *The Shock of the Old*, Edgerton criticizes historians of technology for their preoccupation with technological innovation rather than use.[10] Because technological innovation is geographically and temporally limited, innovation-based narratives can only consider a small part of the world. Yet since every human culture uses some kind of technology, anyplace on earth that people live is a potential setting for a use-based

the present, historians stayed in a historiographical never-neverland in which the historical past receded ever further from the present. See *India After Gandhi: The History of the World's Largest Democracy* (New York: Ecco, 2007), 12–14.

[9] For other scholarship on history of technology in postindependence India, see for example: Ross Bassett, "Aligning India in the Cold War Era: Indian Technical Elites, the Indian Institute of Technology at Kanpur, and Computing in India and the United States," *Technology and Culture* 50 (October 2009): 783–810; Ross Bassett, *The Technological Indian* (Cambridge, MA: Harvard University Press, 2016); David Arnold, *Everyday Technology: Machines and the Making of India's Modernity* (Chicago: University of Chicago Press, 2013); Itty Abraham, *The Making of the Indian Atomic Bomb: Science, Secrecy, and the Postcolonial State* (London: Zed Books, 1998); Daniel Klingensmith, *"One Valley and a Thousand": Dams, Nationalism, and Development* (New Delhi: Oxford University Press, 2007); Satyajit Singh, *Taming the Waters: The Political Economy of Large Dams in India* (Delhi: Oxford University Press, 1997).

[10] David Edgerton, *The Shock of the Old: Technology and Global History Since 1900* (Oxford: Oxford University Press, 2007).

Fig. 1.1 Map of India showing locations of the case studies and other significant places mentioned in the following chapters. Note that while several Indian cities have undergone name changes in recent decades, this book uses the geographical names that were in use in the early independence period. Modern names of cities, where applicable, are given in parentheses on the map (Drawn by the author)

history of technology. One such place is Nehruvian India. With a few exceptions, Nehruvian India was not a site of major technological innovation, but industrial technology was in widespread use there. This book is not a story of how new technologies were developed, but how existing technologies were adopted and adapted to Indian conditions (Fig. 1.1).

Overview of the Book

After this short introduction, the first two chapters of this book comprise an extended introduction, providing background for what follows. Chapter 2 sets the domestic technological scene and shows how India's industrialization had been limited before independence. After independence, Nehru's Congress government tried to correct this shortcoming by launching a series of five-year plans, which would develop India's economy and industrial base and set the nation on the road to autarky and self-reliance. Chapter 3 explains the international context of the Cold War. As the five-year plans were underway, nonaligned India found itself drawn into the struggle between the United States and the Soviet Union, as the rival superpowers raced to develop India through foreign aid. India tried to play the competing sides off each other to get better aid, but without much success.[11]

The following six chapters look at some part of Nehruvian modernization. Each case-study offers insight into how the desire for autarky and nonalignment interacted with the necessity for aid in India's modernization. Chapters 4 and 5 are about technology in the Indian Air Force. In Chapter 4, the IAF begins importing jet fighters at independence and subsequently tries to develop a domestic fighter, the HF-24 Marut, but the results are disappointing. Chapter 5 tells how India's border war with China in 1962 affected military modernization. India received military aid from the western bloc, and the United States, Britain, and Australia participated in a joint air exercise code-named Shiksha, which was staged in North India a year after the war.

The last four full-length chapters are studies of infrastructural projects, which represent a range of approaches to the problem of technology transfer. Chapter 6 narrates the construction of the Saraighat Bridge across the Brahmaputra River in Assam, executed by two Indian firms with limited foreign input. Chapter 7 is about the Umiam Hydroelectric Project in the Khasi Hills of Assam (now Meghalaya), which

[11] David C. Engerman notes that India's playing the Soviet Union and United States off each other was not successful in *The Price of Aid: The Economic Cold War in India* (Cambridge, MA: Harvard University Press, 2018), 2.

was designed in India but financed by the US government and used some imported machinery. Chapters 8 and 9 are both about Tarapur Atomic Power Project, India's first commercial nuclear power station. The American firm General Electric designed the plant, while another American firm, Bechtel, served as the engineering and construction subcontractor. Chapter 8 places Tarapur in its Cold War political context, while Chapter 9 analyzes the labor and technical challenges of the plant's construction.

Arguments

How successful was Nehruvian modernization? Certainly not as successful as the planners had hoped it would be. The five-year plans were unable to make India's economy fully modern and industrial. The socialistic command economy, touted as the way for India to modernize with equity and justice, ultimately failed to deliver the goods. In the face of domestic and international pressure, the Indian government dismantled the command economy in 1991.

There are several reasons why command-economy industrialization failed to live up to the expectations of the planners. One reason was that the plans simply were not realistic. Nature places constraints on humans' attempts to change it.[12] Like the laws of nature, economics also placed limits on development. Thermal and hydro power stations generated electricity for cities and industry, but as the cities grew, the electric grid could not keep up with demand. Even in a country as big as India, there was only so much that the economy could do at a time. By focusing on manufacturing in the early five-year plans, India neglected its agricultural sector, which ordinarily would have been a priority for a large country with an ever-growing population. Instead, India came to depend on foreign food aid.[13]

The Cold War placed its own constraints on Indian development. Aid in general—food or development, eastern bloc or western bloc—was a

[12] Nick Cullather makes a similar point about development projects in Afghanistan during the Cold War, in "Damming Afghanistan: Modernization in a Buffer State," *Journal of American History* 89, no. 2 (September 2002): 512–37.

[13] For food aid and Indian manufacturing, see Nick Cullather, "Hunger and Containment: How India Became 'Important' in US Cold War Strategy," *India Review* 6, no. 2 (April–June 2007): 59–90.

double-edged sword. While aid was supposed to lead to greater self-reliance, it generally led to more aid and therefore more dependence. In the Economic Cold War, aid was a weapon wielded by the superpowers to advance their own self-interest, which rarely aligned perfectly with India's. The United States, for example, hoped that aid would draw India toward the western bloc, while India hoped that by developing it could be better-positioned to remain nonaligned. When India resisted aligning itself with the western bloc, the United States withheld its choicest military technology from India.

Yet while the shortcomings of the Nehruvian command economy are undeniable, its accomplishments should not be overlooked. Despite its failures, the command economy greatly expanded India's industrial capacity. This growth was especially marked in the power industry. Between the beginning of the First Five-Year Plan in 1951 and the launch of the Fourth Plan in 1969, electrical generative capacity in India grew more than fivefold. This power fed newly built factories, and the industrial labor force in the country doubled between 1951 and 1969. Meanwhile, the development of other resources and infrastructure that had begun during the colonial period continued under the command economy. In the same period between the First and Fourth Five-Year Plans, the total area of farmland in the country under irrigation grew by 30 percent. The total route-length of the Indian Railways grew by eight percent and the electrified route-length increased more than sixfold.[14]

Nehruvian development had mixed results. What about the attempts to set up import-substitution industries and indigenize foreign technologies? Here the results were even more ambivalent. Yet how we judge these attempts depends on our understanding of the words "indigenous" and even "Indian." In the strictest nationalistic sense, very few technological indigenization programs were successful—that is, few were able to create

[14] Frank Moraes, ed., *The Indian and Pakistan Year Book and Who's Who 1951* (Bombay: Times of India, 1951), 116, 263, 270, 311; Sham Lal, ed., *The Times of India Directory and Yearbook Including Who's Who 1969* (Bombay: Times of India Press, 1969), 85, 90, 117, 236. The figures for these economic indicators are: electrical generative capacity of just under 2,000 MW (1951) and 10,189 MW (1969); industrial labor force of 2.3 million (1951) and 4.6 million (1969); irrigation farmland area of 49 million acres [20 million hectares] (1951) and 26 million hectares (1969); total railways route-length of 34,022.29 mi [54,753.6 km] (1951) and 59,338.70 km (1969); electrified railways route-length of 236.00 mi [379.8 km] (1951) and 2,836.38 km (1969).

domestic manufacturing facilities fed by domestic supply chains and overseen by Indian experts who knew not only how to make the existing technology but also how to continue its development. This was a tall order, and there were many reasons why, in most cases, it failed, including the inadequacy of domestic supply chains, a lack of domestic markets for the final product, and the reality that the most capable technical experts all too often joined the brain drain to the West.

When I began this project many years ago in graduate school, one of the things I hoped to learn was when and why indigenization had succeeded and failed, in order to draw lessons for the future. Yet when I came to the end of my research and analysis, I was convinced that indigenization is ephemeral and autarky is a mirage. Instances in which indigenization seems to have worked are the exception rather than the rule. Despite nationalists' beliefs to the contrary, the world is big and nation-states are small and artificial and their borders are porous. Just as people and ideas move across borders, so too does technology.

Only in rare instances has a technology been brought across a border and stayed in the new country without receiving additional inputs from outside. Most of the time, technological dialogue persists after technologies cross borders. Historians of technology conceptualize this movement as a process of circulation. Circulation is not merely the unidirectional "diffusion" of technology from the West to the rest, as too long believed by Eurocentric scholars. Rather, it is the movement of technology across boundaries thanks to the agency of parties on both sides.[15]

If we take circulation as a given, then it is hardly surprising or disappointing that indigenization in the nationalistic sense rarely occurred. Yet even if industrial technologies did not become "indigenous" in the nationalistic sense, they certainly could in other respects. Many technologies took on distinctively Indian uses and meanings long before they were produced domestically, if indeed they ever were. The reactors of Tarapur Atomic Power Station, though designed and manufactured in San Jose, California, represented the modernizing ambitions of a former colony, and in this sense they were as Indian as the HF-24 Marut "indigenous" jet fighter, made in Bangalore.

[15] For technology circulation see Asif A. Siddiqi, "Technology in the South Asian Imaginary," *History and Technology* 31, no. 4 (December 2015): 341; Kapil Raj, "Beyond Postcolonialism ... and Postpositivism: Circulation and the Global History of Science," *Isis* 104, no. 2 (June 2013): 344–45.

This book, then, is not a narrative of indigenization in the nationalistic sense. Such a narrative would be narrow, teleological, and quite probably boring. Rather, it is a story of technology circulation and use, of the myriad ways that technologies that originated outside of India's borders can become Indian in terms of use and meaning. Domestic politics, economics, culture, and foreign relations all have a part to play in this story.

In one sense, a technology in India is Indian by virtue of being there and being used there, regardless of its country of origin. Yet just to get to this point, a technology from outside of India has to change. It must change in form or use, or both, to adapt to Indian environments and culture. The history of technology circulation in India is the story of this change.

References

1. Abraham, Itty. *The Making of the Indian Atomic Bomb: Science, Secrecy, and the Postcolonial State*. London: Zed Books, 1998.
2. Arnold, David. *Everyday Technology: Machines and the Making of India's Modernity*. Chicago: University of Chicago Press, 2013.
3. Bassett, Ross. "Aligning India in the Cold War Era: Indian Technical Elites, the Indian Institute of Technology at Kanpur, and Computing in India and the United States." *Technology and Culture* 50, no. 4 (October 2009): 783–810.
4. Bassett, Ross. *The Technological Indian*. Cambridge, MA: Harvard University Press, 2016.
5. Cullather, Nick. "Damming Afghanistan: Modernization in a Buffer State." *Journal of American History* 89 (September 2002): 512–37.
6. Cullather, Nick. "Hunger and Containment: How India Became 'Important' in US Cold War Strategy." *India Review* 6, no. 2 (April–June 2007): 59–90.
7. *Design and Construction Features of Selected Dams in India*. New Delhi: Central Board of Irrigation and Power, 1979.
8. Edgerton, David. *The Shock of the Old: Technology and Global History Since 1900*. Oxford: Oxford University Press, 2007.
9. Engerman, David C. "Development Politics and the Cold War." *Diplomatic History* 41, no. 1 (January 2017): 1–19.
10. Engerman, David C. *The Price of Aid: The Economic Cold War in India*. Cambridge, MA: Harvard University Press, 2018.
11. Guha, Ramachandra. *India After Gandhi: The History of the World's Largest Democracy*. New York: Ecco, 2007.

12. Headrick, Daniel R. *The Tentacles of Progress: Technology Transfer in the Age of Imperialism, 1850–1940*. New York: Oxford University Press, 1988.
13. Klingensmith, Daniel. *"One Valley and a Thousand": Dams, Nationalism, and Development*. New Delhi: Oxford University Press, 2007.
14. Lal, Sham, ed. *The Times of India Directory and Yearbook Including Who's Who 1969*. Bombay: Times of India Press, 1969.
15. Moraes, Frank, ed. *The Indian and Pakistan Year Book and Who's Who 1951*. Bombay: Times of India, 1951.
16. Raj, Kapil. "Beyond Postcolonialism ... and Postpositivism: Circulation and the Global History of Science." *Isis* 104 (June 2013): 337–47.
17. Siddiqi, Asif A. "Technology in the South Asian Imaginary." *History and Technology* 31, no. 4 (December 2015): 341–49.
18. Singh, Satyajit. *Taming the Waters: The Political Economy of Large Dams in India*. Delhi: Oxford University Press, 1997.

CHAPTER 2

The Industrial Revolution in India

The India that gained independence from the British Empire in 1947 was smaller than the India of today. There were only 315 million people in the country then, fewer than there are in the United States now. Many familiar features of twenty-first-century Indian life were absent or rare. There were no three-wheeled autorickshaws, television serials, plastic-wrapped processed food, or bottled water. There were far fewer buses, automobiles, and motorcycles, and much less pollution. Yet much in twenty-first-century India was already there by the middle of the twentieth century. This included not only indigenous cultural artifacts like Hindu temples with peaked roofs or premodern imports like Islamic mosques with rounded domes. It also included modern European imports like railroads, electricity, newspapers, and Indian English.[1]

Something else that would be familiar from 1947 India was the urban–rural divide, which if anything was even more pronounced than it is today. Gandhi was fond of saying that India was a nation of villages rather than towns, and in numerical terms this was certainly the case at independence. Only 15 percent of the country's population lived in urban centers at

[1] Frank Moraes, ed., *The Indian and Pakistan Year Book and Who's Who 1951* (Bombay: Times of India, 1951), 14.

W. A. T. Logan, *A Technological History of Cold-War India, 1947–1969*, Palgrave Studies in the History of Science and Technology,
https://doi.org/10.1007/978-3-030-78767-7_2

independence, while the remaining 85 percent lived in villages, and the majority of them made their livings through agriculture.[2]

Indian agriculture at the end of the Raj was low-capital and labor-intensive. Tractors were practically unknown. Farmers tilled their fields with iron-tipped wooden plows pulled behind a team of hump-backed India oxen. The same oxen also powered pumps that raised water from wells for irrigation. Farmers harvested their fields by hand, cutting stalks of rice and wheat with sickles made by the local blacksmith. To take their crops to market, farmers hitched their oxen to wooden-wheeled bullock-carts, the dominant mode of transportation in India since antiquity. The tools and methods used by many Indian farmers had changed little in hundreds of years.

Though the Indian countryside of 1947 may have appeared timeless to a foreign visitor, it had changed considerably during two centuries of British domination—just as it would continue to change after independence. Colonialism had drawn even isolated farms into the capitalistic world economy, turning subsistence farmers into cash-croppers of cotton or indigo. Railroads crisscrossed the Indian Subcontinent, and huge permanent irrigation works brought water to parched farmland in Punjab.

If the changes brought about by two hundred years of British colonial domination in India were sometimes imperceptible in the country, in the cities they were impossible to miss. Though small in terms of percentage, India's urban population was large in absolute numbers. Nearly fifty million people lived in the cities, a figure comparable to the total population of Britain at the time. In the largest urban area, the twin cities of Calcutta and Howrah in Bengal, clanging streetcars carried office workers to white-columned buildings designed by English architects, which housed the headquarters of businesses owned by British and Indian capitalists. On the other side of the country, in India's second-largest city of Bombay, rows of high-rise apartment blocks lined Marine Drive and looked out over the Arabian Sea.[3]

[2] *Indian and Pakistan Year Book*, 14, 18, 275–76. The official population figures from the 1951 census omitted the tribal people of Assam and the inhabitants of the contested territory of Jammu-Kashmir. For Gandhi's views on the primacy of villages to India, see Divya Joshi, ed., *Gandhiji on Villages* (Mumbai: Gandhi Book Centre, 2002), https://www.mkgandhi.org/ebks/Gandhionvillages.pdf.

[3] *Indian and Pakistan Year Book 1951*, 14.

Mechanized industry, a hallmark of the Industrial Revolution, had first come to India almost a hundred years before independence, but by the late colonial period it still represented a small part of the Indian economy and employed a comparatively small portion of the country's population. The 1931 census identified only 1.2 million laborers employed in large-scale industrial operations, who mostly worked in industries that had been set up by private Indian industrialists receiving varying degrees of support from the colonial government.[4]

It had taken a striking series of changes to get Indian large-scale industry even up to that point. Greater changes were yet to come after independence. This chapter provides an overview of the industrial transformations in India during the late-colonial and early independence periods. Private Indian industrialists established their mechanized industry during the last hundred years of British colonial rule as British imperialists were developing modern infrastructure such as railroads. After independence, the Indian government began trying to modernize the country rapidly through planned development, despite dissenting voices that argued that India should not modernize at all.

A central goal of Nehruvian industrialization was import substitution—setting up factories to produce goods domestically so that they would not have to be imported. After a chronological narrative of Indian industrialization, this chapter does a case-study of one public-sector import-substitution initiative: the heavy electrical machinery industry. The chapter also places import substitution in its broader national and international context of initiatives for economic and technological autarky.

The industrialization of India resembled in some ways the earlier experience of countries in Europe and North America, and yet it was markedly different in other ways as well. The chapter ends with analysis of one of the major differences: how a shortage of resources and an abundance of labor led to a lack of reliance on labor-saving machinery and extensive employment of improvisation and adaptive reuse.

There is no one track to modernization, and India's transformations in the twentieth century illustrate that fact plainly. India is now clearly modern, with its cars, jet airliners, satellite television programs, nuclear power, space program, and mobile phone networks. Yet even as India has modernized, it has displayed a remarkable tolerance for the premodern, as camel-carts, horse-tongas, and animal-powered agriculture continue to exist alongside their more modern counterparts.

[4] *Indian and Pakistan Year Book 1951*, 17.

Industry and the Raj

Long before the industrial revolution began in England, India was a world leader in trade, manufacturing, and export. From ancient times, Mediterranean and European elites had eagerly bought up Indian textiles, ivory, and peacocks. In the early modern period, the forests of central India provided timber for ships built on the western coast for European customers such as the Royal Navy. India's biggest export was cotton goods. Indian calicoes flooded European markets, where domestic producers could not compete with the inexpensive but good-quality foreign products.[5]

The situation turned in the Europeans' favor when British entrepreneurs developed a mechanized textile industry. In the new system of production, the grouping of banks of machines into a single facility, a textile mill, created an economy of scale. The mills required a large labor force to tend the machines (automation would come much later). The advantage of mechanization was that it allowed much higher production levels than traditional methods. Cloth woven in English mills started to outsell Indian calicoes in English markets, and later in India as well. The flood of low-cost English cloth into Indian markets ruined the Indian textile industry. England lacked the climate for growing cotton on a large scale, so much of the cotton processed in English mills had actually been grown in India. In a pattern that would become common in the colonized world, local raw materials (cotton) were exported to be processed in foreign factories, then re-imported as a finished product (cloth), with the value-added profits going to the foreign mill-owners.[6]

Mechanized industry and the factory system of production quickly caught on in Europe and North America, but it would not be until 1856 that the first textile mill was established in India. The colonial government provided official support for the establishment of this first mill in Bombay, but on the whole, the British had an ambivalent policy toward industrialization in India. There was just as much cold indifference as there was active support. It was in the British government's best interest to

[5] Arnold Pacey, *Technology in World Civilization* (Oxford: Basil Blackwell, 1990), 67–9; Dharma Kumar, ed., *The Cambridge Economic History of India* (Cambridge: Cambridge University Press, 1983), 2:19, 24; Andre Gunder Frank, *ReOrient: Global Economy in the Asian Age* (Berkeley: University of California Press, 1998), 198–99.

[6] *Cambridge Economic History of India*, 2:19.

keep India an easily manipulated source of raw materials and a market for finished goods, rather than allow it to become a rising industrial power.[7]

The first Indian captains of industry belonged to classes and communities that had long dominated business and trade in the country. The builders of the mills of Bombay were Parsees, whose ancestors had migrated from Persia after their religion, Zoroastrianism, lost its majority to Islam. The greatest industrial dynasty in India, Tata, was founded by Parsee businessman Jamsetji Tata. After getting his start in textiles, Tata began building a steel mill and industrial township in Jamshedpur, in the eastern Indian state of Bihar, a region with extensive coal and iron ore deposits. American businesses provided technical assistance for the project. Although lukewarm about the undertaking as a whole, the colonial government did provide official assistance in geological surveys, transportation, and import arrangements. Jamsetji died before the plant began production in 1912, but his successors carried on the family name and businesses. The Tata Group now operates in 100 countries and offers such diverse products as automobiles, packaged foods, insurance, telecommunications, and the Taj line of luxury hotels.[8]

In Calcutta, the industrial scene was dominated by the Marwaris, a Rajasthani merchant community. Marwaris expanded their business empires as far east as Assam and as far west as eastern Africa. The biggest name of Marwari business in Calcutta was, and still is, Birla. British capitalists had dominated the jute industry in Calcutta since the 1850s, but Marwari businessmen like G.D. Birla were able to take advantage of the economic instability of the period between the world wars and move in to dominate the market. Other members of the Birla clan grew wealthy from concrete ventures and spent some of their riches establishing a science museum and planetarium in Calcutta and building modern Hindu temples in cities across India.[9]

If the establishment of factories in the colonial era was mainly the prerogative of Indian capitalists, British capitalists and the colonial government, as well as the leaders of some of the semi-independent

[7] *Cambridge Economic History of India*, 2:574.

[8] *Cambridge Economic History of India*, 2:588–89; Gijsbert Oonk, "The Emergence of Indigenous Industrialists in Calcutta, Bombay, and Ahmedabad, 1850–1947," *Business History Review* 88 (Spring 2014): 57–63.

[9] Oonk, "Emergence of Indigenous Industrialists," 48–57.

princely states, played their own significant role by building some western-style infrastructure in India. The most remarkable accomplishment of this period was the construction of the Indian rail network, the largest in any colony, which left India with 24,565 miles of track after Partition. There were also canals to irrigate fields and even a few hydroelectric projects to power cities and factories. It was not much, but it was something. The total electrical generative capacity in India was a little less than 2000 MW at independence, or about 20 percent less than the capacity of the Tennessee Valley Authority in the United States at the time.[10]

Dreams of a New India

At independence, power passed from the British and the princes to the Congress Party, the successor of the Indian National Congress that had led the mass movements against colonial rule. The Congress had been founded in 1885 as a forum for the middle-class Indian elites who felt shut out of the colonial political system. The party continued to be led by elites through independence, but the movement was able to succeed by gaining mass support from the Indian populace. The two most important leaders of the Congress at independence were both English-educated elites, but in their visions for the future of India, they could not have been more different from each other. One, Mohandas K. Gandhi, would be honored as the father of the Indian nation. His birthday, October 2, would be enshrined as one of only three national holidays (the other two marking independence and the adoption of the constitution). But his ideology all but died with him. The other, Jawaharlal Nehru, would lay the groundwork for modern industrial India during his seventeen years as Prime Minister.[11]

Mohandas Karamchand Gandhi was born in Porbandar, a coastal town in Gujarat, in 1869. Although not high-caste (the family name means "grocer"), Gandhi's father held high posts in the administration of the princely state of Porbandar, and he was serving as chief minister at the time of Mohan's birth. The young Gandhi was sent off to study law

[10] *Embassy of India, One Year of Independence* (Washington, DC: Embassy of India, 1948), 31; *Indian and Pakistan Year Book 1951*, 263; *Annual Report of the Tennessee Valley Authority 1948* (Washington, DC: Government Printing Office, 1948), 84.

[11] Bipan Chandra, et al., *India's Struggle for Independence, 1857–1947* (New Delhi: Penguin Books, 1988), 71–3.

in England, where he fell in with such radical intellectual movements as Theosophism and the London Vegetarian Society. After passing the Inner Temple bar and returning briefly to India, he migrated to South Africa, where he spent two decades practicing law and agitating for the rights of Indian immigrants there. Returning to India in 1915, he soon emerged as a leader of the Congress, which he was able to redirect into a mass, nationwide movement that spoke to the interests of the common people. Although he assumed leadership roles in Congress before independence, he refused to have any part in the government of sovereign India. Less than six months after independence, he was dead, killed by an extremist Hindu during an afternoon prayer meeting.[12]

Gandhi developed his philosophy during his formative years in England and South Africa, and he continued to revise and refine his ideas after returning to India. His autobiography, written mostly during his stints as a political prisoner, was fittingly titled *The Story of My Experiments with Truth*. Throughout, Gandhi's thinking was defined by a deep religiosity. His Vaishnavite Hindu upbringing formed the basis of his religion, but he was also influenced by western philosophies he had first encountered in England, notably the radical Christianity of Leo Tolstoy. To Gandhi, the guiding principle for all action should be nonviolence, and courage was the greatest virtue. Nonviolence, wrenched from its Gandhian context, became Gandhi's greatest international legacy. But even though activists in the United States, South Africa, and elsewhere used nonviolence successfully as a tactic, for Gandhi nonviolence inevitably led to other aspects of his philosophy that failed to gain traction either in India or abroad. One such element was his hostile attitude toward modern industrial technology.[13]

In 1908, while sailing between South Africa and England, Gandhi wrote a manuscript that would be published as his first book, *Hind Swaraj, or Indian Home Rule*. In it, he wrote: "I cannot recall a single good point in connection with machinery." He went on to explain that machinery allowed humans to travel at unnatural speeds and overstep the biological limits placed on them by God. Furthermore, through modern industry, a limited number of people were capable of gaining

[12] Louis Fischer, *The Life of Mahatma Gandhi* (New York: Harper & Row, 1950), 10, 13, 25–7, 40.

[13] Fischer, *Life of Mahatma Gandhi*, 194–95.

great wealth and power, while the common people lost what little wealth and power they had as they became an industrial underclass. Factory work was a new form of slavery—slavery to management and slavery to the machines. Rather than follow the western nations and recreate the horrors of Manchester in India, Gandhi hoped that independent India would develop small, localized, village-based economies, such as the hand-spinning of cotton that became the icon of the Congress movement. In this way, he wrote, "We shall save our eyes and money and support Swadeshi [domestic production] and so shall we attain Home Rule."[14]

Gandhi's relationship with machinery was not as simple in reality as his rhetoric suggested. It is ironic, and yet altogether fitting, that he wrote his condemnation of industrial technology while riding aboard a great modern steamship. After committing himself fully to nonviolence and *swadeshi*, he adopted an austere lifestyle, part peasant and part monk. He lived in rural communes that he established and wore a loincloth and shawl made of coarse white fabric that he had spun and woven himself. And yet he carried a pocketwatch, kept close track of his use of time, and sometimes rode around in an automobile. His village crafts program incorporated industrial concepts of work ethic and attention to the clock. Many of Gandhi's faithful benefactors were industrialists. When he died, he was not staying at his commune, but in the New Delhi home of jute mill mogul G.D. Birla.[15]

Like Gandhi, Jawaharlal Nehru was an elite, English-educated lawyer and an ardent nationalist, but the similarities ended there. Twenty years Gandhi's junior, Nehru was born to a wealthy, upper-caste Brahmin family in Allahabad. Although the Nehrus had lived in the north Indian flatlands for centuries, they retained memories of their homeland, Kashmir. (The family name derives from the Hindustani word *nahar*, or canal, because the ancestral Nehrus had lived next to a canal in Kashmir.) Jawaharlal's father, Motilal, was a successful lawyer, and he could afford to keep his wife, son, and two younger daughters in a luxurious mansion

[14] The quotes are from "Machinery," in M.K. Gandhi, *Hind Swaraj, or Indian Home Rule* (1908; repr. Ahmedabad, India: Navajivan, 1938), 80–4. Gyan Prakash discusses the Nehru-Gandhi debate at length in *Another Reason: Science and the Imagination of Modern India* (Princeton, NJ: Princeton University Press, 1999), 201–25.

[15] Ross Bassett, *The Technological Indian* (Cambridge, MA: Harvard University Press, 2016), 14, 90.

named Anand Bhawan, the House of Joy. At the age of fifteen, Jawaharlal shipped off to England to attend an elite public school, Harrow, then study law at Cambridge. When he returned to India, he followed his father into Congress and the independence movement. He served several lengthy sentences as a political prisoner of the Raj, and he passed the time writing three long books: a series of letters to his daughter Indira that were published as *Glimpses of World History*, his *Autobiography*, and his nationalistic manifesto, *The Discovery of India*.[16]

Despite his Kashmiri Pandit heritage, Nehru was never religious. "The spectacle of what is called religion, or at any rate organized religion, in India and elsewhere has filled me with horror," he wrote in his autobiography, "and I have frequently condemned it and wished to make a clean sweep of it. Almost always it seems to stand for blind belief and action, dogma and bigotry, superstition and exploitation, and the preservation of vested interests." Lacking any religious basis, Nehru adopted nonviolence as a tactic rather than a principle, to be employed when expedient and discarded when not. He simplified his dress and lifestyle, but he never followed Gandhi into absolute poverty, nor did he ever fully relinquish his vices (in Gandhi's estimation) of tobacco and meat.[17]

If Nehru had any guiding philosophy as central to his thinking as nonviolence was for Gandhi, it was Enlightenment rationalism: "It is ... with the temper and approach of science, allied to philosophy, and with reverence for all that lies beyond, that we must face life." In holding to his rationalistic philosophy, Nehru dismissed Gandhi's pre-industrial vision for independent India as romantic and impractical. Instead, he wanted India to become a modern, industrial nation. "If technology demands the big machine, as it does today in a large measure," he wrote in *The Discovery of India*, "then the big machine with all its implications and consequences must be accepted." Industrialization was the panacea that would cure all of India's economic ailments. "I am all for tractors and big machinery," he wrote, "and I am convinced that the rapid industrialization of India is essential to relieve the pressure on land, to combat

[16] Frank Moraes, *Jawaharlal Nehru: A Biography* (New York: MacMillan, 1956), 15–6, 22, 28.

[17] Jawaharlal Nehru, *An Autobiography* (1936; repr. New Delhi: Penguin Books, 2004), 390–91; Moraes, *Jawaharlal Nehru*, 213–14.

poverty and raise standards of living, for defense, and a variety of other purposes."[18]

Nehru was not such an idealist to think that machinery and industrialization were unalloyed virtues. He conceded Gandhi's point that far from offering a solution to poverty, industrialization in India had produced capital that had remained in the hands of management. Meanwhile, laborers at some mills lived in such poor conditions that Nehru felt "dazed and full of horror and anger" after visiting their housing on one occasion. Nevertheless, for Nehru the benefits of industrialization outweighed the risks. He was convinced that "the most careful planning and adjustment are necessary if we are to reap the full benefit of industrialization and avoid many of its dangers."[19]

A slogan of the independence movement had been the Hindi word *swadeshi*. Formed of the roots *swa-* (self or own) and *desh* (country), the word in its nationalistic context meant goods of domestic or indigenous manufacture. Under the banner of *swadeshi*, nationalists torched foreign-made cloth in public demonstrations, which they replaced with Indian-made cloth. From 1905 to 1908, *swadeshi* was briefly a movement in its own right, flaring up in opposition to the partition of Bengal. Later, *swadeshi* became a part of the greater nationalist and independence movement. Gandhi and his followers famously wore *khadi*, coarse cotton cloth that they spun and wove themselves. As a subversion of colonial economics, *khadi* was a central symbol of the nationalism of the Indian National Congress. The INC flag even featured a spinning wheel at its center (replaced by the Wheel of Dharma, an ancient Buddhist symbol, after independence).[20]

Promotion of domestic manufacturing also became a pillar of Nehru's postindependence national economic policy, although not in the way it had been promoted before independence. *Khadi* and handicrafts received state support, so they never disappeared, but the majority of resources were directed toward establishing modern mechanized industries. The goal of attaining economic self-reliance, made official in the Third Five-Year Plan, meant that India should not depend on any other nation

[18] Jawaharlal Nehru, *The Discovery of India* (New York: John Day, 1946), 411, 414, 525.

[19] Nehru, *Discovery of India*, 359, 411.

[20] Chandra, et al., *India's Struggle for Independence*, 124–34.

for anything it needed, whether it be foodstuffs, consumer goods, or scientific knowledge. Rather than retreating from the industrial world, as Gandhi had wanted to do, Nehru's India would attempt to join it on equal terms, by expanding the country's agricultural capacity with tractors and fertilizers (part of what would become known as the Green Revolution[21]), building factories to manufacture goods that had previously been imported, and establishing institutions for scientific research and technical education. In this way, India would free itself from the pattern of economic dependence that had been established during the colonial period. With mixed results, programs for import substitution would attempt to manufacture indigenous products ranging from razor blades to sewing machines, bicycles, and fighter jets.[22]

The Five-Year Plans

India's tryst with economic planning began in 1938, when the Indian National Congress established the National Planning Committee. Chaired by Nehru, the NPC consisted of fifteen members, as well as representatives of states that chose to collaborate. The NPC's original mandate was to promote industrialization, but agriculture and social services soon fell under its purview. "The more we thought of this planning business," Nehru wrote, "the vaster it grew in its sweep and range, till it seemed to embrace almost every activity." The committee set goals for production and consumption: a balanced daily diet of 2400–2800 calories for adult workers, 30 yards of fabric consumed per capita per year, housing of 100 square feet per capita, an overall increase in national income of two or three times over ten years.[23]

To chairman Nehru, "The objective of the country as a whole was the attainment, as far as possible, of national self-sufficiency," to steer away from the "whirlpool of economic imperialism." Yet the National Planning Committee would implement none of its plans. Even though some of the states had been under Congress leadership since 1935, the authority

[21] For the Green Revolution, see Nick Cullather, *The Hungry World: America's Cold War Battle against Poverty in Asia* (Cambridge, MA: Harvard University Press, 2013).

[22] B.N. Sinha, *Industrial Geography of India* (Calcutta: World Press, 1972), 14; David Arnold, *Everyday Technology: Machines and the Making of India's Modernity* (Chicago: University of Chicago Press, 2013).

[23] Nehru, *Discovery of India*, 399–402.

of the state administrations was restricted, and the colonial government had little interest in Congress's plans. (The colonial government in fact had its own planning project, on which the Dalit leader B.R. Ambedkar was a committee chairman.) The Congress committee's work came to an abrupt end after the outbreak of World War II and the arrest of Nehru and the other Congress leaders when they refused to cooperate with the war effort.[24]

Planning resumed after the end of the war and independence. From 1950 until his death in 1964, Nehru chaired the Planning Commission, a central government agency responsible for setting targets for national economic growth. The Planning Commission picked up where the National Planning Committee had left off, compiling goals submitted by ministries of the central government and state-level planning committees. Although the checks and balances of the Indian Constitution limited the central government's authority to impose plans on the states, the Planning Commission had the power to sanction financial investment on projects.[25]

The commission's First Five-Year Plan took effect in April 1951. As the introductory chapter of the Plan stated, "The central objective of planning in India at the present stage is to initiate a process of development which will raise living standards and open out to the people new opportunities for a richer and more varied life." The government resolution that established the Planning Commission in March 1950 declared that the goals of planning in India were to provide equal economic opportunity for all citizens, ensure equitable distribution of resources, and guard against "the concentration of wealth and means of production to the common detriment." The mandate stopped short of authorizing wealth redistribution, because this would ruin people's incentive to work. (Wealth redistribution and class struggle, the more radical aspects of Marxism, would take place in West Bengal under a communist state government, but these would not be official policies of the Congress government.) The First

[24] Nehru, *Discovery of India*, 402; Sunila S. Kale, *Electrifying India: Regional Political Economies of Development* (Palo Alto, CA: Stanford University Press, 2014), 32.

[25] Judith M. Brown, *Nehru: A Political Life* (New Haven: Yale University Press, 2003), 205, 240; *Cambridge Economic History of India*, 2:950.

Plan authorized a total expenditure of 23.6 billion rupees ($5.0 billion), mainly on projects that were already in progress.[26]

The Second Five-Year Plan (1956–1961) had a broader scope, with an ambitious program of construction projects that would, it was hoped, lead to a 25 percent increase in national income by 1961. The plan favored industry—especially public-sector industry—over agriculture. (This emphasis reflected the pro-Soviet orientation of statistician Prasanta Chandra Mahalanobis, who outmaneuvered his rivals in the Planning Commission and other government bodies to get his plan adopted.) The total outlay of the Second Five-Year Plan was 48 billion rupees ($10.1 billion), just over double the amount ultimately authorized for the First Plan. Of this sum, the largest portion (28.9 percent) was allocated to transport and communications, two-thirds of which was earmarked for expansion and modernization of the railways. Also receiving significant shares of the plan outlay were irrigation and power (19.0 percent) and industry and mining (18.5 percent). Like the National Planning Committee's unexecuted plans, the Second Plan's mandate extended beyond industrialization and infrastructural development, with programs for modernizing education, expanding access to health services, developing low-income housing, reforming labor policy, and rehabilitating refugees from East and West Pakistan. These social services amounted to 19.7 percent of the plan outlay.[27]

Like the Second Plan, the Third Five-Year Plan (1961–1966) called for industrial expansion targeting a 25 percent increase in national income by the end of the plan. Total plan outlay was 75 billion rupees ($15.8 billion), a 63 percent increase over the 46 billion rupees ultimately spent on the Second Plan. The allocations of funds were similar, with the only major change being a reduction by eight percentage points in the allotment for transport and communications, and an increase for agriculture

[26] Planning Commission, *First Five Year Plan*, http://planningcommission.nic.in/plans/planrel/fiveyr/1st/welcome.html (accessed January 13, 2014); Planning Commission, *Second Five Year Plan*, http://planningcommission.nic.in/plans/planrel/fiveyr/2nd/welcome.html (accessed January 13, 2014); *Cambridge Economic History of India*, 2:954.

[27] Planning Commission, *Second Five Year Plan*, http://planningcommission.nic.in/plans/planrel/fiveyr/2nd/welcome.html (accessed January 13, 2014); *Cambridge Economic History of India*, 2:954; David C. Engerman, *The Price of Aid: The Economic Cold War in India* (Cambridge, MA: Harvard University Press, 2018), 120; Paul Bareau, Roland Bird, and Andrew Shonfield, "India's Second Five-Year Plan," *International Affairs* 33, no. 3 (July 1957): 301–9.

and power to make up the difference. What really distinguished the Third Plan from the second was the vision that animated it. The Third Plan included introductory chapters about perspective or long-term planning over a period of fifteen years or more, a topic that had not received more than a passing mention in the first two plans. According to the Third Plan's expansive introduction, the Indian economy was taking its first step toward becoming "self-reliant and self-generating." As it was, though, the Third Plan would require heavier foreign investment than its predecessors, as a balance of payments crisis of the late 1950s had drained India's foreign exchange reserves.[28]

India's five-year plans, and the Planning Commission that created them, grew out of a widespread twentieth-century belief in technocracy, or rule by technical experts (scientists, engineers, and economists). Proponents of technocracy believed that it was possible for technocrats to engineer away the problems of an economy or society, if they were given full control over the system. The first state to initiate centrally directed industrialization on a large scale was the Soviet Union, although planning was already a popular idea among leaders of less-industrialized countries long before Stalin's First Five-Year Plan began in 1928. It was also popular in the more prosperous and industrialized West; the state-directed development of the Tennessee Valley in the United States was a technocratic enterprise.[29]

Other examples aside, it was the rapid industrialization of the Soviet Union that served as the gold-standard for planning in the mid-twentieth century. Between 1928 and 1940, iron and steel production in the Soviet Union had increased fourfold and electric power generation sevenfold. Soviet workers had built dams, railways, and canals. Not only had industrial capacity expanded, but domestic technical expertise had expanded as well, with 900,000 trained engineers in the country by 1941. These engineers and their successors would develop the Soviet atomic bomb

[28] Planning Commission, *Third Five Year Plan*, http://planningcommission.nic.in/plans/planrel/fiveyr/3rd/welcome.html (accessed January 13, 2014); *Cambridge Economic History of India*, 2:954–55.

[29] Morris Bian argues convincingly in *The Making of the State Enterprise System in Modern China: The Dynamics of Institutional Change* (Cambridge, MA: Harvard University Press, 2005) that Chinese planning under Mao was based on theories formulated by Sun Yat-sen in the early twentieth century; the plans were not mere copies of Stalin's program.

in 1949 and launch the world's first artificial satellite in 1957 and first manned spacecraft in 1961.[30]

Nehru, like other leaders of recently independent nations, was dazzled by the apparent success of the Soviet project, and he wanted to replicate the results in his country. For those without the ideological baggage of the western bloc, it was easy to be impressed by the Soviet Union's accomplishments. Only later in the twentieth century did the inherent shortcomings of the Soviet model became obvious even to the most uncritical observers. Apart from the enormous human toll of dislocation, dispossession, and starvation, Soviet industrialization and collectivization created an undiversified economy that favored big over practical, leading to price crashes and environmental devastation. These same economic weaknesses were also unwittingly built into the Indian Five-Year Plans.[31]

Independence and Autarky

Nehruvian India's promotion of domestic production and import substitution was part of an international trend of pursuing autarky or economic self-reliance in countries across Asia, Africa, and Latin America. Many of the countries chasing autarky had recently been decolonized. All had found themselves serving principally as the donors of resources rather than the earners of value-added profits in the world market.[32]

As early as the late eighteenth century, the American Federalist leader Alexander Hamilton had drawn a link between industrial development and national sovereignty when he advocated industrialization in the fledgling American Republic to reduce dependence on Britain. More than a century later, in the global depression of the 1930s, a wave of national self-sufficiency movements swept Europe, and it was then that the word "autarky" first came into use in the context of economic self-reliance. After World War II, the capitalist countries of western Europe lost interest

[30] David Christian, *Imperial and Soviet Russia: Power, Privilege and the Challenge of Modernity* (New York: St. Martin's Press, 1986), 277, 286.

[31] For critiques of Soviet industrialization, see Loren R. Graham, *The Ghost of the Executed Engineer: Technology and the Fall of the Soviet Union* (Cambridge, MA: Harvard University Press, 1993); Boris Komarov, *The Destruction of Nature in the Soviet Union* (White Plains, NY: M.E. Sharpe, 1980).

[32] David Edgerton, *The Shock of the Old: Technology and Global History Since 1900* (Oxford: Oxford University Press, 2007), 118.

in autarky as they moved toward economic union. But outside Europe, from Indonesia to Egypt to Argentina, autarky experienced a new surge of popularity.[33]

The leaders of these countries were mostly nationalists who had fought to drive European imperialists out of their country but had, at the same time, internalized European critiques of their economies and societies. While claiming to represent their countries' teeming masses, they were elites who had European-style educations. The modernizing elite represented by Nehru and others such as Sukarno of Indonesia, Gamal Abdel Nasser of Egypt, and Julius Nyerere of Tanzania, all wanted to liberate their countries from foreign economic influence and improve the lives of their citizens as a whole.

There were as many strategies for pursuing autarky as there were former colonies, and then some. Industrialization played a major role in the plans of many modernizing elites. India built dams and factories. Egypt poured almost all of its modernizing energy into a single colossus of a dam, the world's largest at the time. Import substitution was a common feature of autarky programs. Indonesia made locomotives to run on the narrow-gauge railways of Java.[34] In Nasser's Egypt, restrictive import regulations virtually mandated domestic manufacture of consumer goods, although even raw materials often still had to be brought in by complicated multi-part barter agreements.[35]

Many countries across Asia, Africa, and Latin America even took the radical step of expropriating foreign assets or seizing them outright. Mexico had led the way by nationalizing its petroleum industry in 1938. Iran tried to do the same thing in 1951. Nasser seized the Suez Canal in 1956. And Fidel Castro's Cuba nationalized all US-owned assets between

[33] Edgerton, *Shock of the Old*, 118; Gordon S. Wood, *Empire of Liberty: A History of the Early Republic, 1789–1815* (Oxford: Oxford University Press, 2009), 100–3. Autarky was also popular in communist countries that were isolated from the outside world and from each other, such as North Korea and Albania.

[34] One of the indigenous Indonesian locomotives, Bima Kunting 3, is on permanent display in a park in front of Benteng Vredeburg, a historic Dutch fort in the central Javan city of Yogyakarta.

[35] Edgerton, *Shock of the Old*, 118. Literary critic Edward Said's father ran an office-supply business in Cairo during Nasser's Arab socialism. His domestically assembled office furniture was made of imported steel. Edward W. Said, *Out of Place: A Memoir* (London: Granta, 2014), 288.

1959 and 1963. The countries whose assets had been seized usually retaliated with boycotts, sanctions, coups, or even military invasions. Yet for nationalists in the nonaligned world, it was often worth it, because taking over foreign assets was a way to stand up to the foreign powers that for so long had exploited their countries.[36]

India in the early independence period still had some foreign-owned businesses, but nothing as powerful or manipulative as the oil industry in Mexico and Iran or the American-owned businesses in Cuba. Nehru's government felt no reason to declare war on foreign capital. (The Janata government of 1977–80, however, did.) Without large foreign businesses to expropriate, Nehru's India would use import substitution as the key to unlock autarky.

Import Substitution and Nehruvian India

Along with planned development, import substitution offered its own tantalizing benefits for the newly independent nation: the saving of foreign exchange by producing products domestically rather than importing them. With import substitution, India could have the best of both worlds—factory-produced products and capital both staying in India. This was what the planners hoped, anyway. As import substitution programs were implemented in practice, they proved to be more complicated matters than simply building factories in India and importing machines to fill them. Whether they lived or died depended on their domestic supply chains and markets.

Both the government and private industry set up import-substitution operations in India in the early independence period. Although the captains of private industry tended to have less of an ideological commitment to autarky and the command economy than their counterparts in the public sector, they recognized that it would be good for their bottom line if they could corner the protected domestic market. The possibilities for import substitution were limitless, but the products that did best

[36] David S. Painter, "Oil and the American Century," *Journal of American History* 99, no. 1 (June 2012): 26, 30–31; Matías F. Travieso-Díaz, "Resolving U.S. Expropriation Claims against Cuba," in *Paths for Cuba: Reforming Communism in Comparative Perspective*, ed. Scott Morgenstern, Jorge Pérez-López (Pittsburgh: University of Pittsburgh Press, 2018), 118.

were those that had a large domestic market, like razor blades, bicycles, and sewing machines.

Although the domestic market for automobiles was not quite so large, the Indian automobile industry was another successful case of import substitution in the private sector. Two companies, Hindustan Motors in Calcutta and Premier Automobiles in Bombay, started mass-production of domestic cars by buying complete tooling for European cars that were being retired. The Morris Oxford became the Hindustan Ambassador and the Fiat 1100 became the Premier Padmini. These cars were produced for decades, and they are still a regular sight on the Indian roads despite the arrival of newer domestic and imported cars.[37]

Like Nehruvian development projects, import-substitution initiatives of the public sector could be impressive, as the full resources of the Indian state were marshaled to build new factories and staff them with qualified personnel. The HF-24 Marut "indigenous" jet fighter (discussed in Chapter 4) was one such public-sector project. Another was the heavy electrical machinery industry, built from the ground up by the Indian government beginning in the mid-1950s.

In 1948, just after independence, the Indian government had set up a committee to study the problem of how to manufacture turbines, generators, transformers, switchgear, and other heavy electrical equipment for use in power stations. The committee's recommendations bore fruit in 1955, when the government established Heavy Electricals (India) Ltd. (HE(I)L) in Bhopal. The balance of payments crisis in the late 1950s provided impetus for the further expansion of the domestic heavy electricals industry. In response, the government established three new facilities in 1960, in Tiruchirapalli, Hyderabad, and Haridwar. These were spun off into a new company, Bharat Heavy Electricals Ltd. (BHEL) in 1964. (In 1971, HE(I)L merged with BHEL.)[38]

In the cities where they were built, the heavy electricals plants were industrial pioneers. Before independence, most of India's heavy industry had been located in a few coastal cities—Bombay, Calcutta, and Madras

[37] R.C. Bhargava and Seetha, *The Maruti Story: How a Public Sector Company Put India on Wheels* (Noida: Collins Business, 2010), 4–5; *The Complete Encyclopedia of Motorcars: 1885 to the Present* (2 nd ed., New York: E.P. Dutton, 1973), 358.

[38] V. Krishnamurthy, "Management of organizational change: The BHEL experience," *Vikalpa* 2, no. 2 (April 1977): 113–19; D.C. Baijal, "March towards self-sufficiency: Power development in India," *Indian and Foreign Review*, February 15, 1966, 10–11.

being the most prominent. After independence, the Planning Commission attempted to spread the benefits of industrialization around, by setting up industrial colonies in previously unindustrialized areas. As M.V. Pylee of the Delhi School of Economics noted in a 1961 article in *Asian Survey*, the decision to locate the first heavy electricals plant in landlocked Bhopal made little economic sense. Because the factory made heavy use of imported materials, it would have made more sense from an economic perspective to build the facility in a coastal location, thereby saving on shipping costs. From a political perspective, though, the decision to locate the plant in Bhopal was perfectly reasonable, because it reflected the Nehruvian desire to industrialize India as a whole, not just the parts of the country that already had industry.[39]

All the heavy electrical plants were located in inland cities that had limited or no mechanized industry by the time of independence. Although the name Bhopal would later become synonymous with the world's worst industrial disaster—the 1984 chemical spill by Union Carbide—the city only became a major industrial center after independence. Tiruchirapalli (or Trichy), Hyderabad, and Haridwar were also industrial latecomers, and located far away from ports. The Heavy Electrical Equipment Plant (HEEP) in Haridwar was built more than 800 miles from the two closest major ports, Calcutta on the east coast and Kandla on the west.

Most of the heavy electrical plants were located far away from the centers of these cities. This was not uncommon in the postindependence industrialization of India, which was largely an extra-urban phenomenon. Factories were built outside of cities, where land was cheaper and there were fewer people to displace—but still close enough to the city that it was easy to run a branch rail line out to the factory. Urban services such as electricity, water, and sanitation might be poor or inadequate, so these did not attract industrialists to locate their projects within city limits. The heavy electricals plants thus had to provide services that the local infrastructure was not equipped to provide. As an example of the transformed industrial geography of Nehruvian India, the Heavy Power Equipment Plant (HPEP) in Hyderabad was not actually located in the city proper,

[39] M.V. Pylee, "Government enterprises in the Indian economy," *Asian Survey* 1, no. 7 (September 1961): 16–26; *Cambridge Economic History of India*, 2:963.

but sixteen miles to the northwest, in an area known as Ramachandrapuram.[40] HPEP was not merely a single building or a cluster of factory halls; it was a miniature city, with a township with 2,307 quarters, a shopping center, a hospital, two primary schools, and a secondary school. The site also featured purpose-built wells and sanitary and storm sewers.[41]

The establishment of the Indian heavy electricals industry used equipment and technical expertise from both eastern- and western-bloc countries. The British firm Associated Electrical Industries served as technical consultant for the establishment of HE(I)L Bhopal. Czechoslovak firms Technoexport, Škodaexport, and ČKD-Praha provided most of the machinery and expertise for the High Pressure Boiler Plant (HPBP) at Tiruchirapalli and HPEP in Hyderabad. HEEP in Hardwar received technical assistance from the Soviet Union. Later, BHEL also signed technical collaboration agreements with ASEA (Sweden), Nuovo Pignone (Italy), and Combustion Engineering (USA).[42]

The donor countries provided both machinery and training for Indian specialists. Hundreds of Indian engineers and technicians traveled abroad for training at the facilities of the collaborating foreign firms. Upon returning to India, the trainees served as factory supervisors and managers. The factories also needed a skilled labor force, but since the heavy electricals industry was new to India, this labor force had to be trained from scratch. To train shop floor workers, management established schools at the plants. Some students entered the training schools with no prior manufacturing experience. At the end of the two-year course, they were ready to enter the regular workforce of the factory.[43]

[40] "Heavy Electrical Project," *India News*, November 8, 1963.

[41] Bharat Heavy Electricals, Ltd., *First Annual Report 1964–65* (New Delhi: Bharat Heavy Electricals Ltd., 1965); Bharat Heavy Electricals, Ltd., *Second Annual Report 1965–66* (New Delhi: Bharat Heavy Electricals Ltd., 1966).

[42] S. Swayambu, "A Decade of Progress in Power Equipment Manufacture," *Indian Journal of Power and River Valley Development* (*IJPRVD*) 21, no. 4 (April 1971): 113–14, 147; S. Sarangapani, "Heavy Electrical Project," *Assam Tribune Magazine*, November 6, 1960; BHEL, *First Annual Report 1964–65*; "Starting Production of Heavy Power Equipment in India," *Czechoslovak Heavy Industry*, May 1966, 20; BHEL, *Second Annual Report 1965–66*; "Bharat Heavy Electricals' maiden profit," *IJPRVD* 21, no. 9 (September 1971): 354–55; "BHEL boiler know-how pact with US firm," *IJPRVD* 21, no. 3 (March 1971): 109; A. Chakravarti, "The Social Profitability of Training Unskilled Workers in the Public Sector in India," *Oxford Economic Papers* 24, no. 1 (March 1972): 114–15.

[43] BHEL, *First Annual Report 1964–65*.

Expertise is essential for import-substitution operations, but expertise alone is not sufficient. By the time the technical collaboration agreements expired and the foreign specialists returned home, HE(I)L and BHEL still relied heavily on imported raw materials and parts. In fiscal year 1969–70, the Trichy plant still imported half of its raw materials, while the Hyderabad and Haridwar facilities imported more than 70 percent of their parts. An article in *Economic and Political Weekly* entitled "BHEL's Wonderland Accounting" criticized the company—which had been established for the purpose of import substitution—for relying so heavily on imports for raw materials and machinery. "Though New Delhi tries to give the impression of being highly selective in permitting import of components by private-sector units," the article complained, "BHEL was allowed import of components ranging from 65 percent to 90 percent in different plants."[44]

Using largely imported parts at first, the HE(I)L and BHEL plants began to manufacture Indian-made electrical machines and install them in power stations around the country. They also produced boilers and other equipment for industrial use. Early plants that used mostly domestic machinery were Ennore Thermal Power Station in Madras State and Giri-Bata Hydroelectric Project in Himachal Pradesh. BHEL would later supply machinery for India's second commercial nuclear plant, Rajasthan Atomic Power Station, and later stages of the Umiam Hydroelectric Project (see Chapter 7). Over time, BHEL would come to dominate the Indian market for heavy electrical machinery, as its products accounted for more than half the total market share.[45]

Yet this success was not assured, and BHEL nearly became a casualty of the failures of Nehruvian modernization. From its beginning, the heavy electricals industry had a captive market in the state electricity boards. In response to the Five-Year Plans, the state electricity boards were actively

[44] Bharat Heavy Electricals, Ltd., *Sixth Annual Report 1969–70* (New Delhi: Bharat Heavy Electricals Ltd., 1970); "BHEL's Wonderland Accounting," *Economic and Political Weekly*, May 16, 1970, 792–93.

[45] Bharat Heavy Electricals, Ltd., *Fourth Annual Report 1967–68* (New Delhi: Bharat Heavy Electricals Ltd., 1968); Y.K. Murthy, "The Giri Hydro-Electric Project," *IJPVRD* 18, no. 7 (July 1968): 259–64; "HEL's Power Projects," *Hindustan Times*, December 27, 1965; Swayambu, "A Decade of Progress in Power Equipment Manufacture," 146; "Bharat Heavy Electricals' Maiden Profit," *IJPRVD* 21, no. 9 (September 1971): 354–55; "About Us," *Bharat Heavy Electricals, Ltd.*, http://www.bhel.com/about.php (accessed October 7, 2014).

expanding their capacities, and thus had high demand for transformers, breakers, boilers, turbines, and generators. HE(I)L and BHEL grew to supply this demand. But when the shortcomings of the Third Five-Year Plan led to the delay and scaling-back of the Fourth Plan, orders for heavy electrical machinery fell far short of expectations for plant management.[46]

It was a do-or-die situation, and rather than dying, BHEL found its salvation in the export market. In 1968, the company had exported its first lot of 930 valves to West Germany. Two years later, BHEL received an order from the National Electricity Board of Malaysia for the delivery and installation of two 60-MW boilers for the Tuanku Ja'afar Power Station in Port Dickson, peninsular Malaysia. With revenues buoyed in part by sales of electrical equipment abroad, BHEL reported its first profitable year in 1971. Subsequently, BHEL exported electrical power equipment to countries across the Indian Ocean region and farther afield, including Iraq, Egypt, and New Zealand.[47]

There is a certain irony here. BHEL was originally established for import substitution and autarky, but not even India could provide the company as big a market as it needed in order to succeed. Instead of allowing India to turn its economy away from the world, BHEL became an exporter and a significant earner of foreign exchange.

No two countries are alike, and the motivations and strategies for import substitution differed from one country to the next. When it came to industrialization, what had worked for England, Germany, or the United States was not guaranteed to work in India. Industrialization in India was limited by constraints on capital and resources. This is not to say that India had neither capital nor resources. The country in fact had plenty of both. Mineral deposits and agricultural products turned into hard currency when exported. The problem was that India did not always have the right resources in the right form. There was also almost never enough capital, especially after the balance of payments crisis of the late

[46] BHEL, *Fourth Annual Report 1967–68*; "Bharat Heavy Electricals' Maiden Profit," 354–55.

[47] Bharat Heavy Electricals Ltd., *Fifth Annual Report 1968–69* (New Delhi: Bharat Heavy Electricals Ltd., 1969); Bharat Heavy Electricals Ltd., *Sixth Annual Report and Accounts 1969–70* (New Delhi: Bharat Heavy Electricals Ltd., 1970); "About Us," *Bharat Heavy Electricals, Ltd.*, http://www.bhel.com/about.php (accessed October 7, 2014).

fifties. To make up for the shortage of capital, India could borrow from abroad (as discussed in Chapter 7) or stretch resources to make do with what was on hand.[48]

Waste and Adaptive Reuse in Indian Technology

Since the birth of mechanized industry in India in the nineteenth century, industrialization had been a process of adapting European machinery and practices to Indian labor and conditions. Even such a well-supported operation as the Tata Iron and Steel plant in Jamshedpur was run at a low capacity so that the refractory linings of the furnace would not burn out and require replacement as often as they would if the plant were run at full capacity. In the early independence period, labor and certain resources substituted for capital and other resources that were lacking in India.[49]

Mechanized labor-saving innovations, which were a hallmark of industrialization in Europe and especially North America, proved not to be very popular in India. If a machine could do the work of twenty men, it was almost always more cost-effective in India just to hire the men than buy the machine, since labor was plentiful and cheap. The logic was reversible too: if a machine would be used in the West for a particular task, then it might be possible to hire a dozen or a hundred laborers to perform a similar task in India.[50]

The most extravagant example of substituting labor for machinery took place in the construction of the Nagarjuna Sagar Dam, ninety-five miles southeast of Hyderabad. Begun in 1955 and completed in 1969, the dam was a combined hydroelectric and irrigation project meant to irrigate the arid Deccan with the waters of the Krishna River. The main dam was constructed of rubble masonry, or locally sourced stones joined by mortar. The lower half of the dam, up to 105 feet above the riverbed, was built by hand. Tens of thousands of laborers working on day and night shifts carried stones and trays full of mortar up ramps that zigzagged up the face of the dam. (Above the 105-foot mark, tower cranes took over and completed the dam.) The spectacle of workers climbing the ramps on the

[48] *Cambridge Economic History of India*, 2:955.

[49] *Cambridge Economic History of India*, 2:589.

[50] *Cambridge Economic History of India*, 2:561.

face of the giant dam was a favorite subject of magazine photographers, who often compared it to the construction of the pyramids in ancient Egypt. At the peak of construction, 60,000 men, women, and children were engaged in the project.[51]

Less photographable than labor-for-machinery schemes like Nagarjuna Sagar, but no less a part of India's quest for autarky, was the substitution of one natural resource for another. India has large supplies of certain natural resources useful for industry, such as coal and iron ore, but other resources are limited or missing entirely. These resources had to be imported, but the Indian government was loath to spend its foreign exchange on importing raw materials when a domestic substitute might work instead.

HE(I)L Bhopal faced this situation as it tried to find sources for the windings of electric power generators. Windings are coils of wire wrapped around magnetic steel cores in the rotor and the stator of a generator, which allow the machine to produce magnetic fields that induce electric currents when the rotor turns. They are typically made of copper. India had deposits of copper in Bihar (now Jharkhand), Rajasthan, and elsewhere, but by the mid-1960s, demand outstripped domestic production by a factor of more than seven. Since India lacked a suitable domestic source of copper, and the price of copper fluctuated widely on the international market, HE(I)L commissioned British consultant J.H. Walker to investigate whether it would be economical to use domestically available aluminum instead of imported copper. Walker concluded that aluminum windings did not represent a significant cost savings if used in place of copper. As he concluded in an article about his research published in *Proceedings of the Institution of Electrical Engineers* in 1967, "the use of aluminum in the stator winding would be justified only in a country where aluminum was indigenous and copper imported and the necessity of saving foreign exchange was important." In other words, it would be justified in a foreign exchange-strapped country such as India.[52]

[51] J.S.R. Mohan Rao, "Nagarjuna Sagar Dam," in *Design and Construction Features of Selected Dams in India* (New Delhi: Central Board of Irrigation and Power, 1979), 77–115; *Masonry Dams in India* (New Delhi: Central Water and Power Commission, 1972), 17–20.

[52] J.H. Walker, "Aluminum Windings for Hydroelectric Generators: A Critical Analysis," *Proceedings of the Institution of Electrical Engineers* 114 (October 1967): 1464–70; Sinha, *Industrial Geography of India*, 218–24.

In addition to labor-for-machinery and resource substitution, a third strategy for autarky and low-cost industrialization was the use of secondhand equipment. Just as secondhand markets for consumer goods give people access to products they could not otherwise afford, buying used is a way for an economy to access machinery for which authorities cannot, or do not want to, expend the capital to buy fresh from the factory. In the early independence period, Indian industry bought secondhand machines or entire production lines. An example of the latter, as mentioned above, was when Hindustan Motors and Premier Automobiles bought tooling from European car lines to use in their own factories in India.

The Indian military also kept itself supplied with used weapons and vehicles. The largest, and probably longest-serving, example of secondhand military hardware in India was INS *Vikrant*, the Indian Navy's first aircraft carrier. Originally named HMS *Hercules*, the ship was one of six *Majestic*-class carriers built by Vickers-Armstrong (Newcastle) for the Royal Navy during World War II. These ships were meant to last only three years or until the end of the war, whichever came sooner. In fact, the war ended before any of the six *Majestic*-class carriers were ready to enter service, so the incomplete vessels were mothballed until the navy could decide their fate. Ultimately, five of the carriers were modified and sold to Commonwealth nations: two to Australia and Canada each, and one to India.

After the Indian Navy purchased *Hercules* in 1957, the Belfast shipyard Harland and Wolff spent four years modifying the ship to accommodate modern carrier aircraft. Among changes made during the retrofit, the ship received an angled flight deck and full air-conditioning. The Indian Navy received the newly commissioned INS *Vikrant* in March 1961. With careful maintenance and periodic refittings and modernizations, the navy kept the ship in service until 1997, more than fifty years after its keel was first laid down and long past the structure's original expected service life of three years. By the time of its retirement, the ship had been replaced by INS *Viraat*, another secondhand British carrier.[53]

[53] Roger Chesneau, *Aircraft Carriers of the World, 1914 to the Present: An Illustrated Encyclopedia* (Annapolis, MD: Naval Institute Press, 1984), 134–51; Norma Friedman, *British Carrier Aviation: The Evolution of the Ships and their Aircraft* (Annapolis, MD: Naval Institute Press, 1988), 242; David Hobbs, *Aircraft Carriers of the Royal and Commonwealth Navies: The Complete Illustrated Encyclopedia from World War I to the Present* (London: Greenhill Books, 1996), 107, 200–1. INS *Vikrant* served as a museum

Conclusion

The early independence period was a time of transition in India, not just politically and socially but also economically and technologically. Where India only had a few mechanized industries before independence—notably steel, jute, and cotton—by twenty years after independence the country had a broader range of industries, many of them established for import substitution. Domestic bicycles and sewing machines replaced foreign-made models. India also now had more and better infrastructure for transportation, agriculture, and electric power generation.

The motivation for all this construction and economic development had been autarky, and yet at the beginning of the Fourth Five-Year Plan in 1969, India was no closer to having a "self-reliant and self-generating" economy (in the words of the Third Plan) than it had been in 1947. India continued to import and export. Indian industrial technology remained inextricably linked to Europe, North America, and Japan. As other former colonies were finding, India needed the world market as much as the world needed India's resources. While technology continued to change outside of India, and foreign economies continued to grow, autarky remained an ever-receding dream.

If other examples are any indication, India's quest for autarky would probably have been a failure in any international context. But as it was, the Cold War further complicated India's quest for autarky. With every conflict or struggle becoming a part of the Cold War (in the eyes of the United States, at least), not even the industrialization of India could escape the clash of superpowers.

References

1. Arnold, David. *Everyday Technology: Machines and the Making of India's Modernity*. Chicago: University of Chicago Press, 2013.
2. Baijal, D.C. "March towards Self-sufficiency: Power Development in India." *Indian and Foreign Review*, February 15, 1966, 10–11.
3. Bareau, Paul, Roland Bird, and Andrew Shonfield. "India's Second Five-Year Plan." *International Affairs* 33, no. 3 (1957): 301–9.
4. Bassett, Ross. *The Technological Indian*. Cambridge, MA: Harvard University Press, 2016.

ship from 2001 to 2012, after which it was scrapped. The Bajaj company subsequently produced commemorative motorcycles made from the aircraft carrier's scrap.

5. Bharat Heavy Electricals, Ltd. *Fifth Annual Report 1968–69*. New Delhi: Bharat Heavy Electricals, Ltd., 1969.
6. Bharat Heavy Electricals, Ltd. *First Annual Report 1964–65*. New Delhi: Bharat Heavy Electricals, Ltd., 1965.
7. Bharat Heavy Electricals, Ltd. *Fourth Annual Report 1967–68*. New Delhi: Bharat Heavy Electricals, Ltd., 1968.
8. Bharat Heavy Electricals, Ltd. *Second Annual Report 1965–66*. New Delhi: Bharat Heavy Electricals, Ltd., 1966.
9. Bharat Heavy Electricals, Ltd. *Sixth Annual Report and Accounts 1969–70*. New Delhi: Bharat Heavy Electricals, Ltd., 1970.
10. Bharat Heavy Electricals, Ltd. *Third Annual Report 1966–67*. New Delhi: Bharat Heavy Electricals, Ltd., 1967.
11. Bhargava, R.C., and Seetha. *The Maruti Story: How a Public Sector Company Put India on Wheels*. Noida: Collins Business, 2010.
12. Brown, Judith M. *Nehru: A Political Life*. New Haven: Yale University Press, 2003.
13. Chakravarti, A. "The Social Profitability of Training Unskilled Workers in the Public Sector in India." *Oxford Economic Papers* 24 (1972): 114–15.
14. Chandra, Bipan, Mridula Mukherjee, and Aditya Mukherjee. *India Since Independence*. New Delhi: Penguin Books, 2008.
15. Christian, David. *Imperial and Soviet Russia: Power, Privilege and the Challenge of Modernity*. New York: St. Martin's Press, 1986.
16. Cullather, Nick. *The Hungry World: America's Cold War Battle against Poverty in Asia*. Cambridge, MA: Harvard University Press, 2013.
17. *Design and Construction Features of Selected Dams in India*. New Delhi: Central Board of Irrigation and Power, 1979.
18. Edgerton, David. *The Shock of the Old: Technology and Global History Since 1900*. Oxford: Oxford University Press, 2007.
19. Engerman, David C. *The Price of Aid: The Economic Cold War in India*. Cambridge, MA: Harvard University Press, 2018.
20. Fischer, Louis. *The Life of Mahatma Gandhi*. New York: Harper & Row, 1950.
21. Frank, Andre Gunder. *ReOrient: Global Economy in the Asian Age*. Berkeley: University of California Press, 1998.
22. Gandhi, M.K. *Hind Swaraj, or Indian Home Rule*. 1908; repr. Ahmedabad, India: Navajivan, 1938.
23. Kale, Sunila S. *Electrifying India: Regional Political Economies of Development*. Palo Alto, CA: Stanford University Press, 2014.
24. Moraes, Frank, ed. *The Indian and Pakistan Year Book and Who's Who 1951*. Bombay: Times of India, 1951.
25. Moraes, Frank. *Jawaharlal Nehru: A Biography*. New York: MacMillan, 1956.

26. Murthy, Y.K. "The Giri Hydro-Electric Project." *IJPVRD* 18 (1968): 259–64.
27. Nehru, Jawaharlal. *The Discovery of India*. New York: John Day, 1946.
28. Oonk, Gijsbert. "The Emergence of Indigenous Industrialists in Calcutta, Bombay, and Ahmedabad, 1850–1947." *Business History Review* 88 (Spring 2014): 43–71.
29. Pacey, Arnold. *Technology in World Civilization*. Oxford: Basil Blackwell, 1990.
30. Painter, David S. "Oil and the American Century." *Journal of American History* 99, no. 1 (2012): 24–39.
31. Planning Commission. *First Five Year Plan*. http://planningcommission.nic.in/plans/planrel/fiveyr/1st/welcome.html (accessed January 13, 2014).
32. Planning Commission. *Second Five Year Plan*. http://planningcommission.nic.in/plans/planrel/fiveyr/2nd/welcome.html (accessed January 13, 2014).
33. Planning Commission. *Third Five Year Plan*. http://planningcommission.nic.in/plans/planrel/fiveyr/3rd/welcome.html (accessed January 13, 2014).
34. Prakash, Gyan. *Another Reason: Science and the Imagination of Modern India*. Princeton, NJ: Princeton University Press, 1999.
35. Pylee, M.V. "Government Enterprises in the Indian Economy." *Asian Survey* 1, no. 7 (1961): 16–26.
36. Raychaudhuri, Tapan, Irfan Habib, Dharma Kumar, and Meghnad Desai. *The Cambridge Economic History of India*. 2 vols. Cambridge: Cambridge University Press, 1982–83.
37. Sinha, B.N. *Industrial Geography of India*. Calcutta: World Press, 1972.
38. "Starting Production of Heavy Power Equipment in India." *Czechoslovak Heavy Industry*, May 1966, 20.
39. Swayambu, S. "A Decade of Progress in Power Equipment Manufacture." *IJPRVD* 21 (April 1971): 113–14, 147.
40. Walker, J.H. "Aluminum Windings for Hydroelectric Generators: A Critical Analysis." *Proceedings of the Institution of Electrical Engineers* 114 (1967): 1464–70.

CHAPTER 3

India and the Cold War

Among the leaders of the Congress Party who led India into independence, Pandit Nehru was easily the most widely traveled. Others had been educated in England or, like Gandhi, had worked abroad. But it was wealthy Nehru who had covered the most miles by 1947, for business and pleasure.

During the seven years he spent as a student in England, Nehru often spent holidays on the Continent, in Berlin or Paris. As an adult and freedom-fighter, Nehru returned to Europe several times in between prison sentences. During these peregrinations in Europe, he became an international figure, an emissary of the Indian independence movement to the world. While his ailing wife Kamala received treatment at sanatoria in Switzerland and Germany, Nehru met with Indian political exiles and attended the Congress of Oppressed Nationalities in Brussels. In the summer of 1937, after Kamala's death in Switzerland, Nehru visited Burma and Malaya, and in 1938 he returned to Europe once more before the outbreak of war. The following year, he flew to Chungking, the wartime capital of China, where he met with Chiang Kai-shek and his American-educated wife Soong Mei-ling.[1]

[1] Jawaharlal Nehru, *An Autobiography* (1936; repr. New Delhi: Penguin Books, 2004), 28, 156, 170–74, 628–32; Jawaharlal Nehru, *The Discovery of India* (New York: John Day, 1946), 26, 33.

W. A. T. Logan, *A Technological History of Cold-War India, 1947–1969*, Palgrave Studies in the History of Science and Technology,
https://doi.org/10.1007/978-3-030-78767-7_3

By 1947, none of the other Congress leaders could claim to have traveled this widely, so Nehru assumed the role of Foreign Minister at independence, which he held concurrently with the office of Prime Minister until his death. In this dual role of chief diplomat and head of government, Nehru was ideally empowered to develop and execute his own distinctive foreign policy doctrine of nonalignment, which would define India's foreign relations for decades and, as an ideology, would reach beyond India's borders.[2]

In this chapter, we turn to India's relationship with the world outside its borders. As much as the planners of Nehruvian India may have wanted to isolate their country's economy from the outside world, the reality was that India needed the world for its resources, capital, machinery, and ideas. This technology, when it came into India, did not come out of a vacuum; it came from an international environment saturated with cold-war politics.

In the context of the Cold War, the outside world was not one but three: the western bloc, led by the United States; the eastern bloc, led by the Soviet Union; and the nonaligned or third world, of which India was itself a leader, a first among equals. Granted, the three worlds were not monoliths. Canada and some of the countries of western Europe chafed against American domination of the western bloc. Eastern European countries were at times unwilling hostages to the Soviet bloc, and when they tried to assert their independence—as Hungary and Czechoslovakia did—the Soviets sent in tanks to keep them in their place. The two largest communist states, the Soviet Union and China, went from being allies in the 1950s to enemies in the 1960s. And within the nonaligned world, countries could hardly agree with each other on anything. The one thing they could agree on was that they had no interest in joining either the eastern or the western blocs.

Nehruvian India was thus indisputably a member of the third world, and this status influenced its relations with countries of all three worlds. Understanding India's relationships with the three worlds is a key to understanding Nehruvian modernization, because cold-war politics affected India's access to the technology it needed and wanted for modernization. This chapter examines India's relations with each of the three worlds, starting with fellow nonaligned nations and continuing with the United States and the Soviet Union.

[2] Judith M. Brown, *Nehru: A Political Life* (New Haven: Yale University Press, 2003), 244.

A major question that this chapter grapples with is why India had strained relations with the United States.[3] On the surface, the two countries seemed like natural allies, as both were liberal democracies with an expressed commitment to secularism, religious freedom, and human rights. The United States poured billions of dollars into the Indian economy during the Cold War, in the form of food aid and grants and loans for development.[4] Yet India was hardly an American ally; it tilted back and forth between the United States and the Soviet Union before finally falling into the Soviet camp in the 1970s.

What had happened? Why did the two countries fail to get along with each other? Economic ideology was clearly a factor. The United States was in favor of capitalism and free markets, while India was developing a socialist economy and chasing after autarky. In addition to ideology, this chapter emphasizes the importance of cultural relations between India and the major powers as an explanation for the rocky relations with the United States. In cultural exchanges, the Soviet Union took India seriously in a way that the United States did not.

Nonalignment in India and Abroad

Nehru developed his philosophy of nonalignment in the early years of the Cold War, when it was obvious that the Grand Alliance between the western Allies and the Soviet Union was broken for good. Before the end of World War II, when he was writing *The Discovery of India* in his cell in Ahmadnagar Fort, it was still too early to tell that the United States and Soviet Union would become such fierce enemies so quickly. Nonalignment was thus not a part of Nehru's vision of an independent India. But he did hint that the Indians were not alone in their

[3] Book-length works that also explore this topic include Robert J. McMahon, *The Cold War on the Periphery: The United States, India, and Pakistan* (New York: Columbia University Press, 1994); Andrew J. Rotter, *Comrades at Odds: The United States and India, 1947–1964* (Ithaca, NY: Cornell University Press, 2000).

[4] By 1966, the United States had contributed a total of $7.6 billion to India in grants and loans. Five years later, that figure had risen to almost $9.9 billion. USAID Mission to India, "The United States Contribution to Indian Development," June 30, 1966, *Development Experience Clearinghouse*, https://pdf.usaid.gov/pdf_docs/PDACQ720.pdf (accessed September 18, 2021); USIS, "Fact Sheet: United States Economic Assistance to India, June 1951–April 1971," Development Experience Clearinghouse, https://pdf.usaid.gov/pdf_docs/PCAAC160.pdf (accessed September 18, 2021).

desire to look after their own concerns rather than pander to the major powers. "Large parts of Asia and Africa consist of awakened, discontented, seething humanity no longer prepared to tolerate existing conditions," he wrote. "Conditions and problems differ greatly in the various countries of Asia, but throughout this vast area, in China and India, in southeast Asia, in western Asia and the Arab world run common threads of sentiment and visible links which hold them together."[5]

As Nehru envisioned the policy, nonalignment meant refusing to take sides in the Cold War. Rather than joining one of the superpower's treaty organizations, nonaligned India would try to cultivate good relations with all countries, regardless of ideology. Like so many other of Nehru's policies, nonalignment grew out of both high-minded internationalism and narrow nationalism. Domestically, nonalignment offered the promise of securing India's independence. Rather than allow itself to be bossed about by one of the great powers in a treaty organization, nonaligned India would exercise the right to make its own decisions in international affairs. Nehru further thought that nonalignment was the key to securing world peace. Although western observers would often confuse the two, nonalignment was not neutralism. As Nehru defined the terms, neutralism meant withdrawing from the international scene—which Nehru had no intention to do—while nonalignment meant the freedom for a nation to make its own decisions in global affairs. Nonaligned India was an active member of the United Nations, frequently taking sides in international disputes. As part of its commitment to the UN, India sent troops for peacekeeping missions, starting in Korea in 1950.[6]

Nonalignment was not the only foreign policy India could have adopted, as illustrated by the trajectory of neighboring Pakistan. Estranged from India by the trauma of Partition, Pakistan followed a different approach to foreign policy, actively seeking out foreign alliances. The smaller and weaker state to emerge from British India, Pakistan needed powerful allies in order to maintain its sovereignty against pressure from India. Immediately after independence, Pakistan began to make overtures to the United States. In 1954, Pakistan gained entry into the American camp by signing a mutual assistance agreement with the

[5] Nehru, *Discovery of India*, 559.

[6] Robert J. McMahon, *The Cold War on the Periphery: The United States, India, and Pakistan* (New York: Columbia University Press, 1994), 37–41; Frank Moraes, *Jawaharlal Nehru: A Biography* (New York: Macmillan, 1956), 472.

United States, which stipulated that Pakistan would receive military aid in exchange for allowing American forces to use strategic airbases and port facilities. (The CIA launched U-2 overflights of the Soviet Union from Peshawar airbase in northwest Pakistan.) Pakistan would go on to join two western-bloc treaty organizations, the South-East Asia Treaty Organization (SEATO) and the Central Treaty Organization (CENTO).[7]

Yet even if Pakistan went its own way, India was far from alone in its belief in nonalignment. As Nehru predicted, the fall of empires in Africa and Asia created new nations whose leaders felt that their territories had more in common with each other than with the major powers. Like Nehru, Nasser of Egypt and Sukarno of Indonesia wanted to modernize their countries and secure economic independence, and they thought the route to autarky lay in nonalignment. Sympathetic western observers dubbed the nonalignment alternative the "Third World," a term that quickly took on economic and even moral senses beyond its original political sense.[8]

In the mid-1950s, it appeared that the "common threads of sentiment" that Nehru had perceived a decade earlier were indeed beginning to bind the nonaligned nations into a unified bloc. In April 1955, delegates from twenty-nine countries met in Indonesia for the seven-day Afro-Asian Conference. Indonesian Prime Minister Ali Sastroamidjojo organized the conference with the support of Nehru and the prime ministers of Pakistan, Burma, and Ceylon. The conference organizers invited delegates from countries in Africa, Asia, and Oceania that had a stated ideological opposition to colonialism. Excluded from the guest list were countries with some complicity in colonialism—as judged by the organizers—including Israel, the Republic of China (Taiwan), Australia, and the Soviet Union.[9]

[7] McMahon, *Cold War on the Periphery*, 142, 183; John Keay, *Midnight's Descendants: South Asia from Partition to the Present Day* (London: William Collins, 2014), 125, 144; Francis Gary Powers and Curt Gentry, *Operation Overflight* (New York: Tower Publications, 1970), 63, 75–80.

[8] B.R. Tomlinson, "What Was the Third World?" *Journal of Contemporary History* 38, no. 2 (April 2003): 307–21. Alfred Sauvy coined the term "third world" in 1952 in an article in *L'Observateur*, in which he explicitly compared the newly independent nations of Africa and Asia with the Third Estate of the French Revolution.

[9] John W. Henderson, et al., *Area Handbook for Indonesia* (Washington, DC: Government Printing Office, 1970), 314. The twenty-nine countries that sent official delegates were Afghanistan, Burma, Cambodia, Ceylon, China, Egypt, Ethiopia, Gold Coast, India, Indonesia, Iran, Iraq, Japan, Jordan, Laos, Lebanon, Liberia, Libya, Nepal, Pakistan,

The delegates met in Bandung, a city in the mountains of Java where Sukarno had once studied engineering. Nehru led his country's delegation to the Bandung Conference, where he hoped to see his vision of nonalignment vindicated. The thirty-seven-year-old Nasser, who had recently assumed control of the Egyptian government and was serving as his country's chief delegate, was reportedly impressed by his meetings with his senior Nehru at the conference, and it was there that he began consciously to develop his own doctrine of nonalignment.[10]

In the end, the Bandung Conference accomplished little. By virtue of their selection, the delegates were naturally willing to endorse a resolution condemning colonialism and racism, but they could agree on little else. The central matter of disagreement was communism. Zhou Enlai, Foreign Minister of the People's Republic of China, represented his country and communism at the conference. Nehru, then courting Chinese goodwill, refused to cast any aspersions on communism. The delegates from Iraq, Ceylon, and the Philippines held no such reservations, and in their speeches at the conference they openly expressed their contempt for communism.[11]

Excluded from the Afro-Asian Conference by virtue of its location, Yugoslavia was the only nonaligned country in Europe. Created after World War I as a unified homeland for the Slavic peoples of the Balkan peninsula, Yugoslavia was a new nation in European terms but older than most other nonaligned nations. The leader of postwar Yugoslavia was Josip Broz Tito, who had risen to power during the war while fighting Germans, Italians, and other resistance groups in his own country. Tito was a communist, and during the war he had been a faithful follower of the Comintern (later Cominform), the Soviet Union's international organization that claimed to speak for all communists worldwide. After the war, Tito launched his country on a reconstruction and industrialization program modeled on the Soviet five-year plans. But in 1947 and 1948, Yugoslav-Soviet relations collapsed, and Stalin expelled Yugoslavia from

Philippines, Saudi Arabia, Sudan, Syria, Thailand, Turkey, North Vietnam, South Vietnam, and Yemen.

[10] Peter Mansfield, *Nasser's Egypt* (rev. ed., Baltimore: Penguin Books, 1969), 98.

[11] Henderson, et al., *Area Handbook for Indonesia*, 315; "Friends of the West Speak Up," *Life*, May 2, 1955, 29–37.

Cominform. Tito stopped trying to copy the Soviets, and instead shifted his focus to decentralization and workers' management of industry.[12]

Cut off from the eastern bloc, Tito developed his own nonalignment doctrine. According to this policy of active coexistence, Yugoslavia would seek positive relations with all other countries, regardless of ideology. By making some pro-West moves, such as closing his southern border to prevent Greek communists from using his country as a base, he attracted economic aid from the United States, Britain, and France. He also normalized relations with the Soviet Union after Stalin's death.[13]

Tito established ties with other nonaligned nations as well. In September 1961, at Tito's invitation, representatives of twenty-four countries met in Belgrade for the first Conference of Non-Aligned Heads of State. Delegates included such familiar figures from Bandung as Nehru, Sukarno, and Nasser, as well as representatives of nations not present at the earlier conference, including Cyprus and Cuba. Like the Bandung Conference, the Belgrade Conference accomplished little. Convened in part to protest the proliferation of nuclear arms, the conference happened to coincide with a resumption of Soviet atomic bomb tests in central Asia. The conference's legacy was that it established a tradition of nonaligned nations meeting in an open forum—a practice that would later come to be known as the Non-Aligned Movement.[14]

As a nonaligned nation straddling the Iron Curtain, Yugoslavia occupied an ambiguous position in Cold War Europe and the world. Open to both East and West, Tito's Yugoslavia was a valuable friend for Nehru's India. Yugoslavia never had one of Europe's major economies, but the nation rebuilt quickly after the war and continued to industrialize after the split with the USSR. Yugoslavia offered India access to European technology without the ideological baggage that attended any imports from aligned nations. In turn, India would pass its technological expertise on to other nonaligned countries such as Egypt.

In the Cold War, India was able to relate to its fellow nonaligned nations on more or less equal terms. But when it came to relations with the eastern and western blocs, and particularly with the superpowers that

[12] Muriel Heppel and Frank B. Singleton, *Yugoslavia* (New York: Frederick A. Praeger, 1961), 174, 187, 194, 199.

[13] Heppel and Singleton, *Yugoslavia*, 203–4.

[14] Tomlinson, "What Was the Third World?" 309, "Sold-Out Belgrade Neutralists," *Life*, September 8, 1961, 50–51.

led them, the relations could not be so equal. In the Cold War, India would have an especially fraught relationship with the United States, the leader of the western bloc and the world's richest and most powerful nation.

India and the United States

In the long years before independence, as he passed his prison sentences writing letters and books and developing his political philosophy, Nehru frequently found occasion to comment on the United States. He had much positive to say about the United States, a country that, with Russia and China, he believed was animated by a "vital energy" absent in other countries of the world. But Nehru was also suspicious of the United States. American isolationism and its anti-democratic adventures in Asia and Latin America provided Nehru ample material for criticism. As an Anglicized, upper-class intellectual, Nehru also looked down on Americans as crude and uncultured—as did many others of his class across the former British Empire.[15]

Nehru's dislike of Americans was mild in comparison with that of others in India and even in his own party. Members of the Planning Commission, who wanted the government to control the economy, tended to be pro-Soviet and anti-American. More visibly, V.K. Krishna Menon, who served as Indian envoy to the United Nations from 1952 to 1960, was an ardent proponent of anti-colonialism, and he directed his fiercest scorn at the United States. In 1957, he famously delivered an eight-hour speech in the UN General Assembly about Pakistan, Kashmir, and the West.[16]

Further to the left of Krishna Menon were the communists. Before independence, the Communist Party of India had been part of a nationalistic coalition with the Congress, but shortly after independence the communists declared that freedom under Congress was a sham. In the first general elections of 1951–1952, the CPI won 23 seats in Parliament,

[15] Sumit Ganguly and Manjeet S. Pardesi, "Explaining Sixty Years of India's Foreign Policy," *India Review* 8, no. 1 (January–March 2009): 5; Andrew J. Rotter, *Comrades at Odds: The United States and India, 1947–1964* (Ithaca, NY: Cornell University Press, 2000), 89–90; Nehru, *Discovery of India*, 44.

[16] Ramachandra Guha, *India After Gandhi: The History of the World's Largest Democracy* (New York: Ecco, 2007), 146–47, 174–75; David C. Engerman, *The Price of Aid: The Economic Cold War in India* (Cambridge, MA: Harvard University Press, 2018), 10.

making it the largest opposition party—although still a distant second to Congress's 364 seats. Communists remained a peripheral force in national politics, but the Communist Party of India established strong regional bases in the south and east.[17]

To the right of Congress and the communists were the commercial interests, with allies in the Finance and Trade ministries in the government. Congress dominated the national conversation, so even the private capitalists had to pay at least lip-service to the party's mission to expand the public sector at the expense of the private. The Bombay Plan, issued by prominent industrialists including J.R.D. Tata in 1944–1945, dutifully expressed support for economic planning. But in spite of the Bombay Plan, most Indian industrialists were understandably more interested in expanding their own businesses than developing the public sector. S.L. Kirloskar, an MIT-educated mechanical engineer and heir of a Maharashtrian business family, came out as a particularly vocal critic of the command economy. With his commitment to the free market, he was a sympathetic figure to the American press, but he was not pro-American so much as he was a global businessman, interested in working with any part of the world, not just within India's artificially constrained command economy.[18]

Even with its plans for autarky, Congress had its own international vision. International engagement was a priority for the Congress government, and even before independence Nehru's provisional government had begun setting up diplomatic missions in foreign capitals. In 1946, the government purchased two turn-of-the-century houses on Massachusetts Avenue in Washington, which were connected to each other and remodeled to serve as the Embassy of India. Another Indian government institution in Washington, the India Supply Mission, was a holdover from the war, when it had served as the colony's representative for lend-lease aid. After the war and independence, ISM stepped into a new role as

[17] John H. Kautsky, "From Proletarianism to Nationalism," *Commentary*, July 1959, 86–90; Shalini Sharma, "'*Yeh azaadi jhooti hai!*' The Shaping of the Opposition in the First Year of the Congress Raj," *Modern Asian Studies* 48, no. 5 (2014): 1358–88.

[18] Sunila S. Kale, *Electrifying India: Regional Political Economies of Development* (Palo Alto, CA: Stanford University Press, 2014), 74; Ross Bassett, *The Technological Indian* (Cambridge, MA: Harvard University Press, 2016), 222–29; Engerman, *The Price of Aid*, 10.

coordinator of development aid and purchaser of parts and materials for India's industrialization.[19]

Nehru made the first of his many trips to the United States in November 1949, just two years after Partition. Although Nehru met with President Truman and addressed both houses of Congress, government-to-government diplomacy was only a part of his agenda. He went to introduce India, and himself as its leader, to the strongest nation in the West. Nehru toured the National Gallery of Art, met New York financiers, watched *South Pacific* on Broadway, addressed students at Columbia and Wellesley, and posed for pictures everywhere—even arm-in-arm with a taxi driver at Boston's Logan Airport. The American press was charmed by the eloquent, well-dressed Nehru; *Life* called him "one of the most likable foreign emissaries who had ever visited this country" (Fig. 3.1).[20]

Nehru's charm and eloquence came as a surprise to the Americans because India played a minimal role in the American imagination at the beginning of the Cold War. Up to that point, the United States had had limited contact with India. American interests in Asia had long resided in China, while India was seen as a wholly British concern. American businesses had not invested much in India, and there were few people of South Asian origin settled in the United States. As Congressman Emmanuel Celler remarked in 1947: "Most Americans still think of India as a land of minarets and performers of the rope trick."[21]

American media had done their part to create and perpetuate these images. At independence, the bestselling book about India by an American author was Katherine Mayo's *Mother India*, published in 1927. American readers devoured Mayo's accounts of India's filth and depravity; by the mid-1950s, over a quarter-million copies had sold in the United States. Even though much in the book was exaggerated at best or fabricated at worst, *Mother India* established India as a place of evil and danger in the American pop culture imagination. Hollywood movies like the

[19] S.C. Aggarwal, *History of the Supply Department, 1939–1946* (New Delhi: Government of India Press, 1947), 35–37. As no other representative of the Indian government existed in the United States at the time, ISM represented India at the inaugural United Nations Conference in San Francisco in 1945.

[20] McMahon, *Cold War on the Periphery*, 56–58; Winthrop Sargeant, "Nehru in the U.S.," *Life*, November 7, 1949, 68–70. Although Nehru made a positive impression in the American press, he clashed with most of the American political leaders he met on his visit.

[21] Quoted in McMahon, *Cold War on the Periphery*, 5.

Fig. 3.1 Prime Minister Nehru with President Kennedy at the Indian leader's arrival in Washington on a November 1961 visit. In the back row are B.K. Nehru, Indian Ambassador to the United States (and the Prime Minister's cousin); US Secretary of State Dean Rusk; and US Vice President Lyndon B. Johnson (*Source* JFK Library, JFKWHP-1961-11-06-A)

prewar *Gunga Din*, set in India but shot in California and featuring no Indian talent, had a similar influence. India and Indians were so far from the experience of the majority of Americans that an African-American, John Roland Redd, was able to pass as an Indian musician, Korla Pandit, on a postwar American television show.[22]

In 1947, India was not much more important to policymakers in Washington than it was to the average American citizen. During the war, India had been a keystone of British strategy. Although more concerned with

[22] Rotter, *Comrades at Odds*, 1–4; Michael Macy, "The Indian Enigma on American TV," *SPAN*, https://span.state.gov/entertainment/indian-enigma-american-tv/20100503 (accessed April 10, 2017).

Europe and the Pacific, the United States deployed forces to India during the war, as the colony offered a route to China after the Burma Road fell to the Japanese. India was also apparently Japan's next target, a supposition confirmed by the Japanese invasion of the princely state of Manipur. About 120,000 American troops deployed to India during the war, where they spent more time building airfields and other strategic facilities than they did fighting. Planes and pilots based at airfields in Assam also shuttled supplies across the Himalaya to China. The China-Burma-India Theater did not get as much coverage in American newspapers and newsreels as the European and Pacific theaters, in part because there were few Allied victories to report until late in the war. Apart from vague memories of the Hump airlifts, the US military experience in wartime India was not enshrined in the American cultural canon in the same way as the battles of Europe and the Pacific.[23]

As soon as the war ended and India was no longer in danger of falling to the Japanese, South Asia became strategically irrelevant to the United States, and the country remained irrelevant in the opening years of the Cold War. The foreign policy strategy of the early Cold War was containment, or encircling the Soviet Bloc with ports and airfields located in nations friendly to the West. On a continental scale, India was not far from the Soviet Union. By air, Delhi was 840 miles from Stalinabad (Dushanbe), capital of the Tajik Soviet Socialist Republic and one of the USSR's southernmost major cities. This was comparable to the distance between Berlin and Leningrad (St. Petersburg). But India would not be significant in the containment strategy, because Nehru and Congress would never have given permission for the United States to operate military bases in the country. Further, Central Asia was not a priority for containment, because the strategy focused on the Cold War flashpoints: eastern Europe, Korea, and southeast Asia. In 1949, the National Security Council concluded that an alliance with India would be a liability rather than an asset.[24]

The attitude of American indifference toward India would not last long. In 1949, the same year the NSC declared its indifference toward

[23] Bassett, *The Technological Indian*, 167–69. For an example of postwar memory of the Hump airlift, see Ernest K. Gann, *Fate Is the Hunter* (New York: Simon & Schuster, 1961), 254–72.

[24] Nick Cullather, "Hunger and Containment: How India Became 'Important' in US Cold War Strategy," *India Review* 6, no. 2 (April–June 2007): 60.

India, Mao's victory in China redrew the lines of the Cold War. Communism was now more widespread than western observers had previously appreciated. To President Truman's advisers, China had succumbed to communism because its people were poor. Communist agitators had exploited poverty with their rhetoric of class warfare and oppression by property-owners. Other countries with large, poor populations could also fall to communism. India was in obvious danger. The Truman administration believed that the way to prevent countries like India from following China's lead would be to strike at the root of the problem, the poverty that had left the door open to radicalism.[25]

In 1951, following a long appropriations debate in Congress, Truman authorized a wheat loan to India. The first shipment of two million tons of American wheat, intended to provide relief for droughts in Bihar and Madras, arrived in Bombay in August. It was the beginning of a long program of food aid to India. Aid expanded immensely in 1954 when President Eisenhower, Truman's successor, signed PL-480, the Agricultural Trade Development and Assistance Act or Food for Peace. Under PL-480, the United States sold wheat and other staples to countries experiencing food insecurity. The US government signed the first PL-480 agreement with India in August 1956. Under the terms of the agreement, the United States sold $362 million worth of American-grown wheat, rice, cotton, tobacco, and dairy products to India.[26]

In the short term, PL-480 was advantageous to the political and economic agendas of both parties of the agreement. The United States was able to dispose of unsaleable surplus wheat, which had been created by New Deal price supports. The Indian government was able to pay for the wheat with rupees rather than foreign exchange reserves. The imported wheat allowed the country to redirect resources from agricultural development to industrialization. The wheat was sold at controlled prices in 50,000 government shops set up around the country, acting

[25] Cullather, "Hunger and Containment," 61.

[26] James C. Hagerty, White House press release, May 4, 1960, S. No. 7, Dwight D. Eisenhower Library papers, NMML; Fact sheet, "India's Participation in Public Law 480," May 4, 1960, ibid.; [Dwight D. Eisenhower], undated draft speech, ibid.; Dwight D. Eisenhower, "Special Message to the Congress on Agriculture," January 29, 1959; Gerhard Peters and John T. Woolley, *The American Presidency Project*, http://www.presidency.ucsb.edu/ws/?pid=11523 (accessed December 8, 2015). Other recipients of PL-480 aid included Brazil, Colombia, Peru, Indonesia, and Pakistan.

as a brake on the inflation that the Second Five-Year Plan's focus on industrialization would otherwise have caused.[27]

It would not be long before this strategy would begin to reveal its holes. With cheap imported grain on the market, domestic producers struggled and failed to compete, leading to a decline in the agricultural sector. India became increasingly reliant on imported grain as its agricultural sector faltered. Thanks in no small part to PL-480, India became dependent on American assistance. When the aid agreement was renewed in 1960, the US government committed to providing $1.3 billion of food in the form of 587 million bushels of wheat and 22 million bags of rice. Although the press release for the renewed agreement portrayed PL-480 as the solution to India's "chronic shortage due to insufficient domestic production," the shortage was in large part caused by the aid.[28]

In December 1959, Eisenhower signaled a growing American interest in India by making the first state visit of any president to the country. After stops in Karachi and Kabul, Eisenhower landed in Delhi on December 9. The president and Nehru rode in an open-backed sedan from Palam airfield to Connaught Place, as throngs of well-wishers standing by the parade route shouted "Eisenhower ki jai!" (Victory to Eisenhower!) While in Delhi, Eisenhower held meetings with Prime Minister Nehru and President Rajendra Prasad. He also presented himself to the Indian people. At a public meeting held on the Ram Lila Maidan near the gates of Old Delhi, more than a half-million people came to see the American president (Fig. 3.2).[29]

As another symbol of growing American interest in India, the Eisenhower administration opened a grand new embassy in Chanakyapuri, a new diplomatic enclave south of Lutyens' Delhi in the capital city. The American diplomatic mission to India had been housed in Bahawalpur House, which the State Department had bought from the Nawab of

[27] Cullather, "Hunger and Containment," 68–71; Nick Cullather, *The Hungry World: America's Cold War Battle Against Poverty in Asia* (Cambridge, MA: Harvard University Press, 2013), 144.

[28] Fact sheet, "India's Participation in Public Law 480," May 4, 1960, S. No. 7, Dwight D. Eisenhower Library papers, NMML; Cullather, "Hunger and Containment," 68–71. The 1960 agreement included $195 million for ocean transport of the food grains to India.

[29] McMahon, *Cold War on the Periphery*, 269–70; "With Ike to the Ends of the Earth: The Triumphant Pilgrimage," *Life*, December 21, 1959, 16–27.

Fig. 3.2 President Eisenhower on his visit to India in December 1959. Here he addresses a huge crowd at Ram Lila Maidan in Delhi (*Source* U.S. Embassy New Delhi Flickr account)

Bahawalpur in 1949 for Rs. 1.6 million ($480,000). During the development of Chanakyapuri, the US government exchanged Bahawalpur House for a 28-acre site between plots that would be taken by the missions of China and France. American architect Edward D. Stone designed the embassy in his signature "stripped-down International Style," with some adaptations for the Indian climate and local materials. Building contractor Tirathram built the facility with the help of a dozen Indian engineers and architects and about 1,200 laborers. The embassy compound included the main Chancery, a consular building, Roosevelt House (the official ambassadorial residence), and a housing complex in the back for the foreign service officers and aid workers who would bring the gospel of capitalism and development to India.[30]

These aid workers were pursuing a new career that represented a different way for the West to understand its role in the rest of the world. "The white man himself is finished in Asia," novelist James Michener had declared in *Life* magazine in 1951. The age of colonialism was over, and Asia was for Asians. But now the white man, and white woman, were being invited back to Asia to serve as development specialists. As agents of the US government, representatives of private non-profits such as the Ford Foundation, or employees of their host country's government, these Americans built dams, designed model villages and cities, and bred more productive strains of staple crops. As Nick Cullather writes, these aid workers' "work remains controversial, and although they are remembered as unsung benefactors or shills for corporate agribusiness, their nation called them to serve."[31]

The election of John F. Kennedy to the presidency of the United States represented another shift in American policy toward India. A World War II naval veteran and son of a powerful Massachusetts senator, the young, photogenic Kennedy won election to the presidency with promises to lead America toward a New Frontier. In his inaugural speech on January 20, 1961, he recast the Cold War as an economic contest rather than an arms race. "To those people in the huts and villages of half the globe struggling to break the bonds of mass misery, we pledge our best efforts to help them help themselves," he declared. "If a free society cannot help the many who are poor, it cannot save the few who are rich." Development, not covert action, would be Kennedy's weapon of choice in the Cold War.[32]

India was a special focus of Kennedy's international development policy for the New Frontier. If India could develop its agriculture, rather than its industry as it was doing under the five-year plans, the Indian people would become prosperous and well-fed—just when millions of people were dying in China from the Great Leap Forward famine. Kennedy

[30] USIS India, "Embassy of the United States of America" (reprint), January 31, 1964, box 277, Master File Copies of Field Publications 1951–1979, RG 306, NARA.

[31] Cullather, *The Hungry World*, 2, 6.

[32] Cullather, "Hunger and Containment," 82; "Inaugural Address, January 20, 1961," *JFK Library*, https://www.jfklibrary.org/Asset-Viewer/BqXIEM9F4024ntFl7SVAjA.aspx (accessed July 18, 2017). Kennedy did, however, authorize plenty of covert operations, especially directed at Castro's Cuba.

appointed several pro-India figures to important positions in his administration: Chester Bowles (who had served as Ambassador to India under President Truman) as Under Secretary of State; John Phillips-Talbot as assistant Secretary of State for Near Eastern and South Asian Affairs; and Robert Komer to the National Security Council.[33]

To succeed career diplomat Ellsworth Bunker, Kennedy made a political appointment of J.K. Galbraith to the Roosevelt House. A Harvard economist who had become a household name for his 1958 book *The Affluent Society*, Galbraith brought the intellectual expertise but not the political skill to oversee India's development as an American satellite. Lanky and very tall, Galbraith towered over his Indian hosts and his own staff. An outspoken opponent of the American alliance with Pakistan, Galbraith endeared himself to the Indians by supporting them in their dispute with Portugal over Goa and all but promising aid that only Congress could authorize. Galbraith was replaced by Chester Bowles in July 1963, who returned to New Delhi for his second term as ambassador. As one foreign service officer recalled, Galbraith was "a big storm that blew through, and then moved on."[34]

To better facilitate the deployment of foreign aid, the Kennedy administration amalgamated PL-480 with another Eisenhower-era initiative, the Development Loan Fund, and other government programs to form a rebranded assistance organization, the United States Agency for International Development (USAID). An executive agency under the Department of State, USAID was dependent on funds released by the legislature. In the politics of 1960s America, Congressional debates about foreign aid appropriations frequently turned contentious. Kennedy's New Frontier Democrats were in favor of aid, but Southern Democrats and opposition Republicans often were not. One of foreign aid's most outspoken and powerful opponents was Democratic Congressman Otto E. Passman of Louisiana, chairman of the House Subcommittee on Foreign Operations Appropriations. Passman believed that aid had the unintended

[33] Cullather, "Hunger and Containment," 82; McMahon, *Cold War on the Periphery*, 275.

[34] McMahon, *Cold War on the Periphery*, 275; Mary Seymour Olmsted, oral history, interview by Charles Stuart Kennedy, April 9, 1992, *Association for Diplomatic Studies*, http://lcweb2.loc.gov/service/mss/mfdip/2004/2004olm02/2004olm02.pdf (accessed June 5, 2014).

Fig. 3.3 US First Lady Jacqueline Kennedy visiting Nehru at Teen Murti Bhawan, his house in New Delhi, March 1962. Also present are Nehru's daughter Indira Gandhi and US Ambassador to India J.K. Galbraith and his wife Kitty (*Source* JFK Library, JFKWHP-1962-03-13-H)

consequence of propping up undemocratic regimes that would be better off collapsing. In cooperation with Republican allies such as Silvio O. Conte of Massachusetts, Passman used his subcommittee position to slash appropriations and intimidate USAID administrators and other aid suppliants.[35]

Conservative American politicians had a variety of reasons for opposing aid in general, and aid to India in particular. One was the belief that American taxes should be spent in support of the American economy, not another country's. In the case of India, Nehru was widely disliked by Republicans and Southern Democrats, who saw him as imperious and aristocratic. Then there was the matter of India and communism. Domestically, the Congress majority tolerated communist and socialist minority parties in India. Internationally, India maintained relations with the Soviet Union by hosting Soviet leaders and accepting Soviet development aid. To the anti-Nehru factions in the US Congress, this was proof that Nehru was a secret Soviet sympathizer and had all but turned his country into a satellite of the USSR. In reality, Nehru's India had stronger economic and technological ties with the United States than with the Soviet Union. But this was the Cold War, and image mattered more than reality. With carefully selected aid, meaningful two-way cultural exchanges, and an expansive (albeit hypocritical) internationalist message, the Soviets had cultivated a more positive image in India than the Americans (Fig. 3.3).[36]

India and the Soviet Union

Nehru first visited the Soviet Union in 1927, at the end of one of his wife Kamala's periods of convalescence in Europe. The occasion was the tenth anniversary of the October Revolution, and he visited at the invitation of the Soviet Society for Cultural Relations with Foreign Countries. Jawaharlal, Kamala, patriarch Motilal, and Jawaharlal's second sister Krishna all accepted the invitation, reaching the Soviet border after a twenty-eight-hour train ride from Berlin. In Moscow, the Nehrus met Soviet President

[35] Billy Hathorn, "Otto Passman, Jerry Huckaby, and Frank Spooner: The Louisiana Fifth Congressional District Campaign of 1976," *Louisiana History* 54 (Summer 2013): 335–36; Hearing, July 24, 1963, in US House Committee on Appropriations, *Foreign Operations Appropriations for 1964* (Washington: Government Printing Office, 1963), 1521–22.

[36] Hearing, August 19, 1963, in US House Committee on Appropriations, *Foreign Operations Appropriations for 1964* (Washington: Government Printing Office, 1963).

Mikhail Kalinin, toured a prison, and took in such standard sights as the Bolshoi Ballet, Lenin's tomb, and the Museum of the Revolution.[37]

Jawaharlal, already a believer in socialism, found much to admire on his visit to the Soviet Union. His father Motilal was not so enchanted. "To my father all such Soviet and collectivist ideas were wholly novel," Nehru wrote in his autobiography. "His whole training had been legal and constitutional, and he could not easily get out of that framework. But he was definitely impressed by what he saw in Moscow."[38]

Even though the Soviet Union had invited private citizen Nehru to visit in 1927, Prime Minister Nehru, and his country, did not get much attention from the Soviets at independence twenty years later. In 1947, the attitude of the Soviet Union toward India was about the same as that of the United States: indifference. Indian and Soviet delegations had met a few times between the end of the war and independence, at the United Nations Conference in San Francisco and the Paris Peace Conference. By 1947, the Soviets had dismissed Nehru and Congress as reactionaries and imperialists. At any rate, the Soviets—like the Americans—were preoccupied with Europe and East Asia. In Europe, the Soviets were busy consolidating their gains from World War II, tightening their control over the territory conquered from the Nazis and now becoming Soviet satellite states. In East Asia, there was China—poised to become Communist, with unknown implications for the Soviets—and Japan, embroiled in a territorial dispute over the Kurile Islands and still technically at war with Russia. Presently concerned with securing its own borders and regional sphere of influence, the Soviet Union had no interest in expanding further into the world.[39]

India and the Soviet Union exchanged ambassadors in 1947, but it was not until 1949 that Stalin even met the Indian ambassador, Nehru's first sister Vijaya Lakshmi Pandit, as she was rotating out of her post. In the early 1950s, relations between the two countries improved slightly. In the summer of 1951, Indo-Soviet cultural exchanges began with the visit of an unofficial Indian delegation to the USSR. The following year, a Soviet delegation returned the visit. Soviet delegates also attended the

[37] Moraes, *Jawaharlal Nehru*, 117–19.

[38] Nehru, *An Autobiography*, 174.

[39] Bhabhani Sen Gupta, *The Fulcrum of Asia: Relations Among China, India, Pakistan, and the USSR* (New York: Pegasus, 1970), 45–51.

first Indian international film festival and an international industries fair in Bombay.[40]

Stalin's death in March 1953 gave Khrushchev and Bulganin, the new leaders of the Soviet Union, the opportunity to reinvent the country's domestic and foreign policy. Nikita Khrushchev, a stout, balding former general of the Russian army in the Great Patriotic War, was First Secretary of the Communist Party of the Soviet Union. This had been the third-highest office in the Soviet Union during Stalin's time, but with skillful maneuvering Khrushchev was able to make his office the most powerful in the government. In official state functions, Premier Nikolai Bulganin formally represented the USSR, but it was Khrushchev who wielded real power. Born in eastern Ukraine to a peasant family, the poorly educated, blustering Khrushchev made an unlikely cultural ambassador for Russia, but he nevertheless played an important role in opening the Soviet Union to the nonaligned world.[41]

An early indicator of the change in approach to India came in September 1953 with the posting of a new Soviet ambassador to New Delhi, Mikhail Menshikov, a foreign trade specialist. Within months of his arrival at the post, Menshikov had negotiated a five-year Indo-Soviet trade pact. Then in 1955, the leaders of India and the Soviet Union each made state visits to the other country. In June 1955, almost six years after visiting Truman at the White House, Nehru made his first visit to the Soviet Union since becoming Prime Minister. Agriculture and industry dominated the Prime Minister's sixteen-day itinerary. In Moscow, Nehru and his daughter Indira Gandhi rode the metro and visited the USSR Agricultural Exhibition and Stalin Auto Works. In Stalingrad, they saw a hydroelectric project and tractor plant. They also visited newly opened farmlands in the Altai, a machine factory in Sverdlovsk, metal works in Leningrad, and the steel plant at Magnitogorsk.[42]

Apart from his tours, Nehru met with Soviet leaders and talked with them at length about internationalism, nonalignment, and Mahatma Gandhi. In a joint statement issued at the end of the visit, India and the

[40] Sen Gupta, *Fulcrum of Asia*, 65–66.

[41] Sen Gupta, *Fulcrum of Asia*, 75; William Taubman, *Khrushchev: The Man and His Era* (New York: Norton, 2003), xi, xix, 312, 354.

[42] *Jawaharlal Nehru in the Soviet Union* (Moscow: State Fine Arts Publishing House, 1955); Sen Gupta, *Fulcrum of Asia*, 75.

Soviet Union declared that neither country would enter into a coalition against the other. As Bhabhani Sen Gupta wrote in 1970, "This was the beginning of nonaligned India's alignment with the Soviet Union."[43]

Later that year, Khrushchev and Bulganin returned the favor of Nehru's visit by taking a tour of southern Asia, visiting Burma, Afghanistan, and India. In India, and to a lesser extent the other two countries, the Soviet leaders offered development aid. Earlier in 1955, the Soviets had already concluded an agreement to build India's first public-sector steel plant at Bhilai in Madhya Pradesh (now Chhattisgarh). During their visit to India, Khrushchev and Bulganin offered vastly expanded aid. Over the next half-decade, the Soviets offered to help build up such Indian industries as petroleum, pharmaceuticals, and heavy electrical machinery. Nehru's Congress Party accepted the aid with smug satisfaction. This was the fruit of nonalignment: India could receive assistance from both sides of the Iron Curtain, whereas Pakistan, currently marching in lockstep with the Americans, was not visited by Khrushchev and Bulganin, and did not receive any Soviet aid offers.[44]

India's own communist party received no respect from the Soviets. Aid to the Indian government would only strengthen the Congress Party and weaken the communists, and the Soviets understood this. It was in fact in the Soviets' best interest for the Congress government to stay in power. Congress was a known quantity; the Indian communists were not. Even though there was a pro-Soviet wing of the party that claimed to be aligned with the Cominform, there was also a pro-Chinese wing. There was no guarantee that the communists would remain loyal to the USSR should they come to power in India. It was just as likely that they might take China's side (after the Sino-Soviet split was formalized in 1961), or even form their own communist bloc of one.[45]

As the 1950s progressed, India and the USSR developed ever more political and ideological common ground. The Indian government abolished *zamindari*, the traditional land-tenure system. In 1956, the legislature adopted the Industrial Policy Resolution, which reserved large parts of the Indian economy for the public sector. The next year, the Congress

[43] Sen Gupta, *Fulcrum of Asia*, 76.

[44] Adam B. Ulam, *Expansion and Coexistence: The History of Soviet Foreign Policy, 1917–67* (New York: Praeger Publishers, 1968), 561–62; Sen Gupta, *Fulcrum of Asia*, 77–78.

[45] Ulam, *Expansion and Coexistence*, 710.

Party's annual session declared socialism to be a national goal. In Kerala, communists won state elections and were installed in power with the approval of the central government. In the United Nations in 1956, India voted against a US-led resolution calling for a plebiscite in Hungary after the Soviet invasion of the country. Even though Nehru privately disapproved of Soviet actions in Hungary, he voted against the resolution because he had always opposed such a plebiscite in Kashmir.[46]

But as the 1950s neared their end, rifts were starting to develop between the Soviet Union and India. The Soviets started to suspect that India was not properly revolutionary. In April 1959, the central government deposed the communist government of Kerala. Also in 1959, conservative business, landlord, and princely interests founded the Swatantra Party, which the Soviets condemned as reactionary. It was at this time that India was tilting toward the United States. Just as Indo-Soviet relations made the Americans uneasy, the Soviets were worried about the strengthening bonds between India and the United States. The enormous crowds that turned out for Eisenhower's visit to New Delhi particularly concerned the Soviets. By the end of the decade, some Soviet intellectuals had started to think that capitalism in India was already too strong, and India had missed its chance to become fully socialist.[47]

For the first half of the 1960s, the Soviet Union took a step back from India. Economic aid and cultural exchanges continued, and India became a customer of Soviet fighter planes for the first time in 1962. But the Soviets did not attempt to engage India in the same way that they had in the late 1950s. When China attacked India in October 1962, India had to turn to the West for aid, because the USSR offered none.

It was after Nehru's death and Khrushchev's ouster, both in 1964, that the Soviet Union, now led by Leonid Brezhnev and Alexei Kosygin, began once more to play an active role in South Asia. In the interest of keeping Congress in power, the Soviets offered to broker a peace settlement between India and Pakistan after the two countries' war in 1965. The belligerents met in the Soviet city of Tashkent in the early days of January 1966, with Premier Kosygin mediating. The resulting Tashkent Resolution failed to ensure lasting peace in South Asia, because India and Pakistan were at war again within six years. But the Soviets' show of

[46] Sen Gupta, *Fulcrum of Asia*, 80–84.

[47] Sen Gupta, *Fulcrum of Asia*, 86–90.

peacemaking served their interests in South Asia. The Congress government stayed in power, even after Prime Minister Shastri's sudden death during the Tashkent conference. As for Pakistan, the Soviets' intervention checked the country's drift toward China. The Soviets had begun playing politics in South Asia again, and they were playing well.[48]

Cultural Exchanges Between the Superpowers and India

As the USSR was playing politics to India, it was also wooing the Indian people through cultural exchanges. Both superpowers were in fact reaching out to the nonaligned world at this time. Yet even though the United States was probably trying harder and certainly investing more resources, the Soviet Union ended up doing a better job marketing itself and its ideology to the nonaligned world.

In the decades after World War II, the United States government aggressively promoted the expansion of American business interests abroad. In Europe, this cultural influence was nearly unidirectional. Along with postwar reconstruction aid came American fashion, chain stores, movies, and teen angst. In later decades, the aggressive exporting of American culture and values would become a genuinely global phenomenon, a major part of globalization.[49]

A half-century later, by the early twenty-first century, evidence of American culture had become omnipresent in India and impossible to miss, from superhero movies in urban megaplexes to New York Yankees caps for sale at village markets. The youth, at least those with exposure to foreign media and the leisure to think about such things, idealized America and looked on it as a promised land. But all this was decades in the future in the days of Nehru, Eisenhower, Kennedy, and Khrushchev. If the common Indian man or woman idealized any country in the early Cold War, it was most likely to be the Soviet Union.

Why did the Soviet Union occupy such a prominent role in the Indian imagination? For the common man and woman, cultural interchange

[48] Ulam, *Expansion and Coexistence*, 715.

[49] Victoria De Grazia, *Irresistible Empire: America's Advance Through Twentieth-Century Europe* (Cambridge, MA: Belknap, 2006), 3–6.

between the two countries mattered most. Significantly, the cultural interchange was two-way. The Soviets took as they gave. Not only did Soviet celebrities like cosmonauts visit India, India returned the favor by sending artists and musicians to the Soviet Union. For the masses, no medium had as wide a reach and profound impact for cultural interchange as film. Indian directors produced films that played to packed cinema halls in the Soviet Union. Raj Kapoor's noir masterpiece *Awaara* was released in the USSR in 1954, the year after Stalin's death and three years after the film's original release in India. In 1954 alone, Soviet citizens bought over 60 million tickets for the movie. In addition to viewing *Awaara* and other pictures, Russian audiences also eagerly bought up commemorative film merchandise. Some Soviet citizens were even lucky enough to catch glimpses of the great stars on their visits to film festivals in Moscow or Tashkent.[50]

At this point, Americans were still reading *Mother India* and falling for Korla Pandit on television. Indian culture would have its day in the sun in the United States, a decade after *Awaara* caused a sensation in the Soviet Union. The American hippie movement, formed in reaction to the Vietnam War and the overconsumptive, materialistic society that supported it, would adopt elements of Indian culture as a way of subverting the affluent society. This was not without precedent. For more than a hundred years, western counterculture had romanticized the East as the antithesis of aggressive, destructive industrial society. Transcendentalists like Thoreau and Emerson read the *Bhagavad Gita* in the mid-nineteenth century, and the founders of Theosophism, Madame Blavatsky and Colonel Olcott, traveled to India on a spiritual quest in 1879. In the imaginations of many western romantics, India, Nepal, Tibet, and the Himalaya were nearly synonymous terms, as they were all parts of the spiritual East.[51]

After the appalling industrialized slaughter of World War I, there was a surge of popularity for the East in western counterculture. Tibet was particularly popular, as it was a place supposedly free from the taint of

[50] Sudha Rajagopalan, "Emblematic of the Thaw: Early Indian Films in Soviet Cinemas," *South Asian Popular Culture* 4, no. 2 (October 2006): 83–100; Alexander Lipkov, "India's Bollywood in Russia," trans. Thomas J. Mathew, *India International Centre Quarterly* 21, nos. 2/3 (Summer–Monsoon 1994): 185–94; "Rūsīantariksh yatriyoṁ kā śāndār nāgarik abhinandan," *Hindustān*, November 11, 1963.

[51] Mark Liechty, "Building the Road to Kathmandu: Notes on the History of Tourism in Nepal," *Himalaya* 25, no. 1 (2005): 20.

western colonialism and industry. In the early 1920s, British mountaineers received permission from the Tibetan government to launch three expeditions to Mt. Everest from the north, as the approach from the south through Nepal was strictly closed to foreigners. The official accounts of these expeditions were widely published and read in the English-speaking world. The expeditions may have inspired American author James Hilton to write the novel *Lost Horizon* (1933), which further popularized the image of the uncolonized, uncorrupted Himalaya.[52]

By the mid-1960s, Tibet was closed to westerners but Nepal and India were wide open. Young, disaffected westerners began traveling in large numbers to South Asia, driving vans, taking buses, or thumbing rides on the overland Hippie Trail through Iran, Afghanistan, and Pakistan. Many of these vagrants had taken to the road because they felt out-of-place in consumeristic America or Britain. Members of this lost generation also flocked to the cafes of Paris, dive hotels of Istanbul, and even caves on the island of Crete, where they squatted for free.[53]

Travelers set out on the Hippie Trail not only because they wanted to get out of America or Britain; they also wanted to get to India or Nepal, the mystical East. (Continental Europeans, who started traveling the Hippie Trail in large numbers later in the sixties, were less interested in India itself than they were in escaping Europe.) The Americans and Britons who drove their rusting vans across the Khyber Pass were attracted by a century's worth of western romanticizing of the East.[54]

As westerners traveled to India, products, ideas, aesthetics, and people also traveled in the opposite direction. Clothing with Indian patterns and cuts appeared in stores from London to Los Angeles. The Beatles and Moody Blues used sitars in their studio albums. Indian holy men, now able to settle in the United States thanks to immigration law reform in 1946 and 1965, set up temples and meditation centers in major cities, attracting curious youths in droves. Indian mysticism had had a tiny following in the United States since the 1890s, but it was only in the 1960s that it became anything close to mainstream.[55]

[52] Liechty, "Building the Road to Kathmandu," 21.

[53] Liechty, "Building the Road to Kathmandu," 22; Thomas Thompson, "Crete: A Stop in the New Odyssey," *Life*, July 19, 1968, 20–28.

[54] Liechty, "Building the Road to Kathmandu," 22.

[55] "Year of the Guru," *Life*, February 9, 1968, 53–59.

The differences between the Soviet and American engagements with Indian culture were fundamental. Superficially, the Raj Kapoor films that Soviet audiences consumed were just movies. But these movies addressed real issues of class, poverty, and privilege—and with their working-class heroes and capitalist villains, they found eager audiences among the Soviet masses. The American hippie movement also addressed real issues like the Vietnam War, but it did so by looking back on an idealized South Asian past while making little attempt to understand the present. If the majority of Indians consumed less than the majority of Americans at the time, it was because the people had less money, not because they were otherworldly. Indian newspapers of the sixties were as full of advertisements for consumer products as American papers.

Soviet mainstream culture engaged with contemporary India. American counterculture romanticized India. For Indians who knew what was happening, it seemed that the Soviet Union was taking their country seriously, but the United States was not.

Conclusion

Nehru's vision of nonalignment proved to be less practical than originally anticipated. From independence, it was only a matter of time before the great powers would start jockeying for influence in India. The Indian government enjoyed playing the two sides off each other for a while, but the balance was precarious, and at length once-nonaligned India tumbled into the Soviet camp.

For its part, the United States was too implicated by the twin evils of colonialism and racism for India ever to accept it as an ally. Unlike Britain, the United States had never controlled much foreign territory directly, but it made up for this with obsessive meddling in Latin America and China. In the Cold War, the United States justified its interventions as upholding democracy in the face of communism, but in reality the interventions protected American business interests and all too often deposed democratically elected leaders. As for racism, the systematic oppression of America's black minority made headlines across the nonaligned world.[56]

[56] For international coverage of the American civil rights movement, see Mary L. Dudziak, *Cold War Civil Rights: Race and the Image of American Democracy* (Princeton, NJ: Princeton University Press, 2000).

Russia was hypocritical in its own way, holding up the banner of anti-colonialism with one hand while suppressing dissent in its eastern European empire with its other. Somehow, Russia's hypocrisy seemed not to matter to India. It could be that India, prejudiced in its own way, did not mind the oppression of Europeans in Europe. In any event, India was dazzled by the Soviet Union's internationalist rhetoric. In his life, Nehru was able to keep India teetering unsteadily between the eastern and western blocs, but in the years after his death, the balance shifted decisively in favor of the USSR. In 1971, seven years after Nehru's death, India signed a twenty-year mutual defense treaty with the Soviet Union.

Even as India tilted away from the West, the United States continued to pour aid into India, and by the year of the signing of the Indo-Soviet pact, total American aid to date had been nearly $9.9 billion. This was more than the Soviet Union provided, and indeed the United States' massive capitalist economy offered far more disposable income than the USSR could ever have mustered. The American aid left its mark on the Indian landscape, but it mattered little politically. "Soviet economic aid to India is less than one-fifth the magnitude of United States aid," American columnist Thomas F. Brady complained in the *New York Times* in 1965, "but dollar for dollar or ruble for ruble the Soviet Union gets about five times as much credit."[57]

In the Cold War, India was able to exploit its ambiguous diplomatic position to gain access to technology from both the eastern and western blocs. This was as India hoped it would be. Yet in the fraught political landscape of the Cold War, India could not get all it wanted. As we will see in the next chapter, India was never able to procure high-performance American fighter jets because it was nonaligned while its bitter rival Pakistan was an American ally. Because of international and regional politics, certain technologies remained always out of reach for India. The Cold War could be a curse to India as well as a blessing.

[57] *United States Economic Assistance to India, June 1951–April 1971* (New Delhi: United States Information Service, 1971), 40–41; Thomas F. Brady, "Soviet's India Aid Fifth That of U.S.," *New York Times*, February 21, 1965.

References

1. Aggarwal, S.C. *History of the Supply Department, 1939–1946*. New Delhi: Government of India Press, 1947.
2. Bassett, Ross. *The Technological Indian*. Cambridge, MA: Harvard University Press, 2016.
3. Brown, Judith M. *Nehru: A Political Life*. New Haven: Yale University Press, 2003.
4. Cullather, Nick. "Hunger and Containment: How India Became 'Important' in US Cold War Strategy." *India Review* 6, no. 2 (April–June 2007): 59–90.
5. Cullather, Nick. *The Hungry World: America's Cold War Battle Against Poverty in Asia*. Cambridge, MA: Harvard University Press, 2013.
6. De Grazia, Victoria. *Irresistible Empire: America's Advance Through Twentieth-Century Europe*. Cambridge, MA: Belknap, 2006.
7. Engerman, David C. *The Price of Aid: The Economic Cold War in India*. Cambridge, MA: Harvard University Press, 2018.
8. Ganguly, Sumit, and Manjeet S. Pardesi. "Explaining Sixty Years of India's Foreign Policy." *India Review* 8, no. 1 (January–March 2009): 4–19.
9. Guha, Ramachandra. *India After Gandhi: The History of the World's Largest Democracy*. New York: Ecco, 2007.
10. Hathorn, Billy. "Otto Passman, Jerry Huckaby, and Frank Spooner: The Louisiana Fifth Congressional District Campaign of 1976." *Louisiana History* 54 (Summer 2013): 332–55.
11. Henderson, John W., et al. *Area Handbook for Indonesia*. Washington, DC: Government Printing Office, 1970.
12. Heppel, Muriel, and Frank B. Singleton. *Yugoslavia*. New York: Frederick A. Praeger, 1961.
13. *Jawaharlal Nehru in the Soviet Union*. Moscow: State Fine Arts Publishing House, 1955.
14. Kale, Sunila S. *Electrifying India: Regional Political Economies of Development*. Palo Alto, CA: Stanford University Press, 2014.
15. Kautsky, John H. "From Proletarianism to Nationalism." *Commentary*, July 1959, 86–90.
16. Keay, John. *Midnight's Descendants: South Asia from Partition to the Present Day*. London: William Collins, 2014.
17. Liechty, Mark. "Building the Road to Kathmandu: Notes on the History of Tourism in Nepal." *Himalaya* 25, no. 1 (2005): 19–28.
18. Lipkov, Alexander. "India's Bollywood in Russia." Translated Thomas J. Mathew. *India International Centre* Quarterly 21, nos. 2/3 (Summer–Monsoon 1994): 185–94.
19. Macy, Michael. "The Indian Enigma on American TV." *SPAN*. https://span.state.gov/entertainment/indian-enigma-american-tv/20100503.
20. Mansfield, Peter. *Nasser's Egypt*. Rev. ed. Baltimore: Penguin Books, 1969.

21. McMahon, Robert J. *The Cold War on the Periphery: The United States, India, and Pakistan*. New York: Columbia University Press, 1994.
22. Moraes, Frank. *Jawaharlal Nehru: A Biography*. New York: Macmillan, 1956.
23. Nehru, Jawaharlal. *An Autobiography*. 1936. Reprint, New Delhi: Penguin Books, 2004.
24. Nehru, Jawaharlal. *The Discovery of India*. New York: John Day, 1946.
25. Powers, Francis Gary, and Curt Gentry. *Operation Overflight*. New York: Tower Publications, 1970.
26. Rajagopalan, Sudha "Emblematic of the Thaw: Early Indian Films in Soviet Cinemas." *South Asian Popular Culture* 4, no. 2 (2006): 83–100.
27. Rotter, Andrew J. *Comrades at Odds: The United States and India, 1947–1964*. Ithaca, NY: Cornell University Press, 2000.
28. Sargeant, Winthrop. "Nehru in the U.S." *Life*, November 7, 1949, 68–70.
29. Sen Gupta, Bhabhani. *The Fulcrum of Asia: Relations Among China, India, Pakistan, and the USSR*. New York: Pegasus, 1970.
30. Sharma, Shalini. "'Yeh azaadi jhooti hai!' The Shaping of the Opposition in the First Year of the Congress Raj." *Modern Asian Studies* 48, no. 5 (2014): 1358–88.
31. Taubman, William. *Khrushchev: The Man and His Era*. New York: Norton, 2003.
32. Thompson, Thomas. "Crete: A Stop in the New Odyssey." *Life*, July 19, 1968, 20–28.
33. Tomlinson, B. R. "What Was the Third World?" *Journal of Contemporary History* 38, no. 2 (2003): 307–21.
34. Ulam, Adam B. *Expansion and Coexistence: The History of Soviet Foreign Policy, 1917–67*. New York: Praeger Publishers, 1968.

CHAPTER 4

Modernizing the Indian Air Force: Imported Fighters and the HF-24 Marut, 1947–1968

August 15, 1947. India becomes a self-governing dominion of the British Empire. The new nation is fêted: there are parades and fireworks, politicians make speeches, the national flag is officially raised in the capital and major cities for the first time.

It was a time when, in the words of Prime Minister Nehru, an age ended, and the soul of a nation, long suppressed, found utterance.[1] Yet concurrent with the eloquent rhetoric and flag-waving, a great demographic and political catastrophe was underway in the form of the Partition of the Indian subcontinent into mutually-antagonistic India and Pakistan.

Partition divided communities, disrupted trade routes, and unhinged an avalanche of communal violence and ethnic cleansing—one of the bloodiest episodes of an exceptionally violent century. Partition opened wounds that continue to fester in the twenty-first century. When the Radcliffe Commission drew the new international border from its bureaucratic offices in New Delhi, its members considered only population statistics. Muslim-majority districts were assigned to Pakistan, while other districts fell on the Indian side of the border. The Commission did not

[1] Jawaharlal Nehru, speech before the Constituent Assembly, August 14, 1947, quoted in Frank Moraes, *Jawaharlal Nehru: A Biography* (New York: Macmillan, 1956), 2.

W. A. T. Logan, *A Technological History of Cold-War India, 1947–1969*, Palgrave Studies in the History of Science and Technology,
https://doi.org/10.1007/978-3-030-78767-7_4

conduct field surveys of any of the border areas, and as a result the border haphazardly divided infrastructure such as irrigation works and rail lines.[2]

Partition also ripped the erstwhile colonial administration into two pieces. Harried civil servants divided the contents of libraries and archives, packing up records relevant to Pakistan and shipping them off to the new temporary capital at Karachi. The armed forces were also divided, as the result of a series of contentious meetings by the Armed Forces Reconstitution Committee. The Indian Army and Royal Indian Navy were apportioned between the two dominions, with about two-thirds falling to India's lot and the remaining one-third to Pakistan.[3]

The Royal Indian Air Force likewise fell under the decolonizing surgeon's scalpel. Greatly expanded during World War II, the air force now boasted ten squadrons. As for the army and navy, the air force was also divided by the Armed Forces Reconstitution Committee. Within the committee, representatives of India and Pakistan presented competing claims to sizable shares of the air force. India prevailed with its argument that it deserved a greater share than Pakistan, since it had the larger land area, and the personnel of the air force was 80 percent non-Muslim. India accordingly received eight of the ten squadrons.[4]

These squadrons were immediately pressed into service during the humanitarian crisis of Partition. As long columns of Hindu and Sikh refugees streamed out of West Pakistan into India, Royal Indian Air Force planes airdropped medical supplies and *chapattis* (flatbread) cooked by volunteers in Delhi. Planes also patrolled the road ahead, watching for the armed bands that had been slaughtering refugee caravans.[5]

As the divided Punjab continued to be convulsed by communal violence, war broke out between the newly divided nations, and the RIAF did some fighting of its own. Barely two months after Partition, India and Pakistan went to war over Kashmir, a Muslim-majority princely state with

[2] Yasmin Khan, *The Great Partition: The Making of India and Pakistan* (2007; repr. New Delhi: Penguin, 2013), 126.

[3] "India Gets Its Freedom," *Life*, August 18, 1947, 28–29; Frank Moraes, ed., *The Indian and Pakistan Year Book and Who's Who 1951* (Bombay: Times of India, 1951), 97, 707.

[4] Syed Shabbir Hussain and M. Tariq Qureshi, *History of the Pakistan Air Force, 1947–1982* (Karachi: PAF Press, 1982), 18–24.

[5] Khan, *The Great Partition*, 162; "Aircraft to Aid Refugees," *Times of India* (*TOI*), September 11, 1947; "Punjab Refugees on Move Again," *TOI*, September 19, 1947.

a Hindu maharaja. The maharaja, Hari Singh, hesitated over whether to accede to one of the colonial successor states or perhaps maintain independence as a Himalayan kingdom. When Pakistani militants crossed into Kashmir to try to force the issue, Hari Singh belatedly sent an offer of accession to New Delhi. And so it was war between India and Pakistan, two countries that just months earlier had been one and the same. Indian forces moved into Kashmir to hold the line against the militants from the west. RIAF Spitfires entered the fray, strafing militants on the road to Srinagar. "The raiders were reported to have been seen from the air, scurrying for cover in all directions, leaving behind their trucks and heavy equipment, some of which were badly damaged," reported the *Times of India*.[6]

The fighting ended with a UN-brokered ceasefire defining the Line of Control as the de facto international boundary—although both nations continued to claim all of Kashmir as their own. India and Pakistan would go to war over Kashmir every so often, with low-level conflict and saber-rattling continuing almost incessantly to a greater or lesser degree ever since.

Let us take a step back from the conflict over Kashmir to consider how this struggle has been witness to technological changes. In the first Kashmir war, the Royal Indian Air Force flew propeller-driven Supermarine Spitfires left over from World War II, and Pakistan brought supplies to its ground forces with Douglas Dakotas. In the next Indo-Pakistan War, in 1965, both sides had jets of the Cold War. By the time of the most recent large-scale conflict, in 1999, both sides had recently demonstrated their capability to build and detonate nuclear weapons.

The Kargil War of 1999 is outside the scope of this book, but an earlier part of the trajectory of technological change—the transition from propellers to jets and finally to supersonic jets—is a part of the larger picture of modernization and indigenization programs that is the theme of this book. In the early decades of independence, the Indian government's autarkical policies extended to the military, and industrial works were accordingly set up to produce armaments as import substitutes.

The most ambitious and highest-profile example of arms import substitution in the early independence period was the HF-24 Marut, a single-seat jet fighter developed in Bangalore at Hindustan Aeronautics

[6] "Aircraft Support for Troops Fighting Invaders," *TOI*, October 30, 1947; "Indian Troops on Offensive," *TOI*, October 31, 1947; "Pattan Cleared of Raiders," *TOI*, November 4, 1947.

Ltd. (HAL) by a joint German-Indian team. The HF-24 was envisioned as an import replacement for fighters bought from Britain and France. Entering squadron service in 1964 after a long gestation in the halls of HAL, the HF-24 was deployed as a ground-attack fighter-bomber because it was underpowered for its original role of interceptor. Produced in limited quantities and performing below its designers' expectations, the HF-24 failed to save the Indian government from the need to continue importing foreign fighters.

The modernization of the Indian Air Force in the early independence period shows how autarky clashed with strategic requirements, as these values reflected competing understandings of modernization. Definitions of progress and modernity, and the relative importance of civilian and military technology, were constructed, contested, and subject to change. In a 1956 speech before the Lok Sabha, the lower house of the Indian parliament, Prime Minister Nehru argued that a country's strength was based on its industry, not its military. "The Five-Year Plan is the defence plan of the country," he said. "What else is it? Because, defence does not consist in people going about marching up and down the road with guns and other weapons. Defence consists today in a country being industrially prepared for producing the goods and equipment of defence." Make industry strong, Nehru argued, and a strong military will naturally follow.[7]

Not everyone in India accepted Nehru's argument that militarization should be subordinate to industrialization. The IAF leadership spent much of the fifties and sixties longing to have American supersonic fighters, which remained out of reach because of nonalignment and the politics of the Cold War. The Indian interest in American fighters peaked after November 1963, when a squadron of supersonic F-100D Super Sabres of the US Tactical Air Command visited India to take part in a joint air defense exercise codenamed Shiksha (as described in more detail in the next chapter). The IAF leadership was preoccupied with supersonic American fighters in part because they were jealous of the American types flown by the Pakistan Air Force and were anxious to maintain strategic

[7] "The Plan Is the Country's Defence," in *Jawaharlal Nehru's Speeches* (Delhi: Ministry of Information and Broadcasting, 1958), 3, 41; Abraham, *Making of the Indian Atomic Bomb*, 11; Nick Cullather, "Damming Afghanistan: Modernization in a Buffer State," *Journal of American History* 89, no. 2 (September 2002): 513.

parity. But their interest went deeper than that. The IAF also wanted American fighters because the service's identity was based on technology.[8]

The US Army Air Corps in the 1930s had insisted that they needed the bigger and supposedly higher-tech four-engine B-17 Flying Fortress bomber rather than the smaller but equally practical B-18 Bolo. In much the same way, the IAF leadership in the sixties wanted only what they perceived to be the best and most sophisticated technology. In 1963, the IAF was hardly bereft of supersonic aircraft, as the service was already operating the MiG-21 from the Soviet Union. The MiG-21 was a simple, lightweight design that made up for its limited range and armament with speed and maneuverability. By comparison, the American fighters that the IAF leadership wanted were heavier, more expensive, and much more complicated. The IAF equated complexity with technological sophistication, and therefore its leaders wished they could have American rather than Soviet fighters.[9]

This chapter is focused on just one aspect of air force modernization: the acquisition of aircraft, in particular jet-powered fighters. The leaders of the Indian Air Force pursued other forms of modernization in the early decades of independence as well, including reorganizing the service's structure, developing domestic training facilities, and sending officers abroad for advanced training. None of these other reforms attracted as much attention or inspired as much commentary as the aircraft. Even though David Edgerton is correct in his assertion that a technology's visibility is no proof of its significance, visibility does make a subject more accessible. Secondary source materials are a valuable asset when studying the Indian Air Force, because its official records remain classified. There are many articles and books about the aircraft of the IAF, but very few about organizational reforms.[10]

[8] History of the 354th Tactical Fighter Wing (TAC), July–December 1963, Call # K-Wg-354-HI, IRIS # N0728, in the USAF Collection, AFHRA; David R. Devereux, "The Sino-Indian War of 1962 in Anglo-American Relations," *Journal of Contemporary History* 44, no. 1 (2009): 86; "Himalayas Barrier to Radar," *The Times*, November 20, 1963; "Shiksha Improves IAF Interception Techniques," *Hindustan Times* (*HT*), November 17, 1963.

[9] Bill Gunston, *Mikoyan MiG-21* (London: Osprey, 1986), 9, 68; Timothy Moy, "Transforming Technology in the Army Air Corps, 1920–1940: Technology, Politics, and Culture for Strategic Bombing," in *The Airplane in American Culture*, ed. Dominick A. Pisano (Ann Arbor: University of Michigan Press, 2003), 322–23.

[10] Ved P. Bhasin, "The Indian Air Force," *India News*, April 9, 1965; Jasjit Singh, *The Icon: Marshal of the Indian Air Force Arjan Singh, DFC—An Authorised Biography* (New

Although one jet fighter was designed and produced in India in the early independence period, India was not a center of aeronautical innovation. The majority of the Indian Air Force's aircraft, fighters or otherwise, were imported or at most license-produced. The Indian Air Force never acquired American jet fighters, but it did fly some multi-engine transports from the United States. It also flew fighters from Britain, France, Russia, and India itself. Technology circulated not just from the West to India, but also between Russia, Egypt, Argentina, and India.

This chapter explores that circulation in all its messy complexity, from the importation of British and French jets to the HF-24 project, MiG-21 deal, and repeated unsuccessful attempts to procure American supersonic fighters. In the Cold War, nonaligned India could not purchase American fighters comparable to those used by its enemy Pakistan, an American ally. Blocked from getting American fighters, and unable to defend itself fully with Indian-made fighters for technical and economic reasons, India turned to Russia, which would prove to be a dependable supplier of arms.

The Indian Air Force Before Independence

Like the modern Indian Army and Navy, the Indian Air Force originated during the colonial period. Military traditions had existed in India long before the East India Company and the Raj, and the British founders of the armed forces drew on these traditions and incorporated some of them (such as rank structure) into the Indian Army and Navy. The Air Force, on the other hand, was something new. The *vimānas* (flying chariots) of the Sanskrit epics notwithstanding, India had no air force tradition until the British brought theirs from Europe after World War I. The Royal Air Force began operating in India in 1921, performing a strategic mission that had not existed until a few years earlier.[11]

That mission was air control. General Hugh Trenchard, Chief of the Air Staff for the Royal Air Force, promoted air control as a means

Delhi: KW Publishers, 2009), 124–26. For visibility and significance, see David Edgerton, *The Shock of the Old: Technology and Global History Since 1900* (Oxford: Oxford University Press, 2007), 1–4.

[11] For precolonial military traditions, see Atul Chandra Roy, *A History of Mughal Navy and Naval Warfares* (Calcutta: World Press, 1972); Raj Kumar Phul, *Armies of the Great Mughals, 1526–1707* ([n.p.]: Oriental Publishers & Distributors, 1978).

of cheaply policing isolated, uncooperative parts of the British Empire. Taking the place of the expensive, land-based punitive expeditions of the nineteenth century, twentieth-century air control missions were comparatively cheap and logistically uncomplicated. Aircraft flew political representatives to villages for negotiations, exhortations, and ultimatums; if these failed, the air force bombed village water supplies and other strategic targets, in theory pacifying the villagers and forcing them to submit to British control. In the Indian Empire, the most troublesome and isolated territory was the North-West Frontier on the border with Afghanistan, which became the focus of air control beginning in 1925.[12]

Air control provided a justification for the existence of the newly independent Royal Air Force. It also provided a mission for the Indian Air Force upon its establishment in 1932. The Indian Air Force was founded as part of post-Great War colonial cost-cutting. The Indian Sandhurst Committee, chaired by Andrew Skeene and including among its members the nationalist leader Motilal Nehru, studied the question of defense and costs. The committee's final report, submitted on November 14, 1926, recommended partial Indianization of the armed forces in order to shift the cost of Indian defense to India. Nationalists also favored Indianization, because it offered them the prospect of greater autonomy and control. The creation of an Indian air force, funded by domestic revenues, was one of the outcomes of the committee's recommendations. The first aircraft flight of the IAF, formed in April 1933, consisted of four Westland Wapiti IIA biplanes, six Indian flying officers, and nineteen other ranks. On their first assignment, the pilots policed the isolated border regions of Waziristan.[13]

[12] David E. Omissi, *Air Power and Colonial Control: The Royal Air Force, 1919–1939* (Manchester: Manchester University Press, 1990), 8–9; Michael Paris, "Air Power and Imperial Defence 1880–1919," *Journal of Contemporary History* 24, no. 2 (April 1989): 209; Phillip Meilinger, "Trenchard and 'Morale Bombing': The Evolution of Royal Air Force Doctrine before World War II," *Journal of Military History* 60 (April 1996): 253–54.

[13] Peter W. Gray, "The Myths of Air Control and the Realities of Imperial Policing," *Aerospace Power Journal* 15, no. 3 (fall 2001): 27; Pushpindar Singh, *History of Aviation in India: Spanning the Century of Flight* (New Delhi: Society for Aerospace Studies, 2003), 112; S.N. Goyal, "The Indian Air Force: It's History, Achievements, and Aims," *Indian Aviation*, June 1952, 179; Bhasin, "The Indian Air Force"; A. Berriedale Keith, "Notes on Imperial Constitutional Law," *Journal of Comparative Legislation and International Law* 10, no. 1 (1928): 119–20; Elisabeth Mariko Leake, "British India Versus the

The Indian Air Force was at first entirely oriented toward British technology. Pilots trained at the Royal Air Force College in Cranwell, and the planes they flew were manufactured by the Westland Aircraft Works in Somerset. The mission of air control was also a British invention. With no indigenous airpower tradition to draw from, the IAF was born as an Anglicized, westernized institution.[14]

The IAF started small and, for the rest of the thirties, built slowly. It was not until 1938 that the first aircraft flight was joined by two others to make No. 1 Squadron. Then in 1939, as part of the military buildup in response to India's entry into World War II, the IAF began to expand rapidly. From having only Wapiti biplanes at the outbreak of the war, within a few years there were seven squadrons of Hurricane fighters and two of Vengeance dive-bombers, as well as one transport squadron equipped with twin-engined Dakotas. The expanded IAF served on the Burma front and, in 1941, assumed full responsibility for air policing on the northwest frontier. Because of its distinguished service in the war, the IAF became the Royal Indian Air Force in 1945, a title it held until India made its definitive break with the British Crown by adopting a republican constitution in 1950.[15]

In the military buildup during the war, the IAF began to diversify its technology from British to American sources. The air force's Vengeance dive-bombers were designed and manufactured in the United States by Vultee Aircraft and sent to India through lend-lease aid. During the war, the IAF also used American-made Lockheed Hudson bombers, North American Texan trainers, and Fairchild PT-26 Cornell trainers. The Dakota transports that entered service just after the war were likewise American products, designed and manufactured in California. When the IAF adopted American planes for the first time in World War II, it was one sign of many that India's technological connections with the United States were growing while those with Britain were in decline.[16]

A legacy of wartime American involvement in Indian aviation was Hindustan Aircraft Ltd., established in 1940. The founder was Seth

British Empire: The Indian Army and an Impasse in Imperial Defence, Circa 1919–39," *Modern Asian Studies* 48, no. 1 (January 2014), 3–7.

[14] Bhasin, "The Indian Air Force."

[15] Bhasin, "The Indian Air Force"; Goyal, "The Indian Air Force," 179.

[16] Ross Bassett, *The Technological Indian* (Cambridge, MA: Harvard University Press, 2016), 167–69; Pushpindar Singh, *Aircraft of the Indian Air Force, 1933–73* (New Delhi: English Book Store, 1974), 40–42, 47–48, 51–56, 69–73.

Walchand Hirachand, an Indian industrialist who had already formed shipping and construction companies. (His construction company, HCC, appears in Chapter 6.) Before founding HAL, Hirachand had been interested in starting an automobile manufacturing plant in India. On a trip to the United States, he encountered aviation entrepreneur William Douglas Pawley, who convinced him to focus on aircraft manufacturing instead.[17]

After returning to India, Hirachand set up a factory in the city of Bangalore, then part of the princely state of Mysore. Twenty-two American specialists traveled to India to oversee the establishment of the plant. The Maharaja of Mysore and the Government of India both owned fifty percent shares in the operation. At the beginning of the war, HAL assembled some military aircraft for use in India, including Curtiss P-36 Hawk fighters and Vultee V-12 attack bombers. In December 1942, the US Army Air Forces assumed control of the facility and began using the plant for overhaul of its own aircraft operating in the China-Burma-India Theater. All American military plane types could be serviced at HAL, including flying boats, which were flown in and out of a water-filled channel on the factory grounds.[18]

The Air Force Looks West

With the end of the war came the double uncertainties of demobilization and decolonization. About 11,000 Indian troops joined the Allied occupation forces in Japan, including members of the RIAF 4th Squadron. When the last of these troops returned to India in October 1947, after a deployment of eighteen months, they returned to a subcontinent divided. Just days before fighting broke out in Kashmir, the former occupation forces split up and joined the armed forces of either the dominion of India or Pakistan.[19]

Popular histories of the Indian Air Force tend to exaggerate the effect of Partition on the service's fighting capability. It is true that two squadrons joined the Pakistan Air Force and the airfields used for air control—many of the force's best—ended up in Pakistani territory. But the RIAF retained eight squadrons and shortly took over some of the

[17] Jasjit Singh, *Indian Aircraft Industry* (New Delhi: KW Publishers, 2011), 30, 44–48.

[18] Hindustan Aeronautics Limited, *Annual Report 1964–65* (Bombay: [n.p.], 1965); "American Makes Planes in India," *Life*, March 22, 1943, 30–32.

[19] "Indian Troops Quit Japan," *TOI*, October 11, 1947; "Indian Occupation Forces in Japan: Arrival in Madras," *TOI*, October 27, 1947.

Fig. 4.1 An ex-IAF B-24 liberator, on display at the Pima Air and Space Museum in Tucson, Arizona, in 2011 (Photo by the author)

RAF airfields in its territory. Although resources were limited, the RIAF was able to meet its immediate strategic requirements even at this stage.[20]

One way the RIAF met its needs was by repurposing hardware left behind by the Allies after the war. On one occasion, India even performed adaptive reuse to induct new aircraft into the fleet. When the British and American armed forces left India after World War II, they abandoned nearly a hundred Consolidated B-24 Liberator bombers at the maintenance depot in Kanpur. In 1947, the RIAF commissioned the newly nationalized HAL to refurbish as many of these aircraft as possible. By November 1948, the first six Liberators were ready for service. HAL ultimately put forty-two Liberators back into service (Fig. 4.1). These planes served on coastal patrol duty, performed a photo-mapping of Mt. Everest after the first successful summiting in 1953, and participated in Operation Vijay, the Indian annexation of Goa in 1961.[21]

Repurposing leftover Allied aircraft could only get the air force so far, and it would not be long before India would start looking to import new aircraft. In 1948, the RIAF inducted its first new fighters since independence—and its first jets—the de Havilland Vampire. Four other NATO-bloc jet fighters followed in the 1950s. From England came the

[20] "R.A.F. Personnel to Be Replaced," *TOI*, May 14, 1947; Bhasin, "The Indian Air Force."

[21] Singh, *Aircraft of the IAF*, 78–80; Kapil Bhargava, "India's Reclaimed Bombers," *Bharat Rakshak*, updated June 13, 2017, http://www.bharat-rakshak.com/IAF/aircraft/past/927-b24.html; "Everest Portfolio by the Indian Air Force," *Mountain World* (1954): 17–18; Attachment to letter, R.J. Rebello to Charles E. Johnson, February 19, 1965, S. No. 18, Ambassador to USA subject files, papers of B.K. Nehru, NMML.

Hawker Hunter in 1957 and the Folland Gnat in 1958. From France, the air force imported the Dassault Ouragon (named "Toofani" in Indian service) in 1953 and the Dassault Mystère IV in 1957.[22]

India's immediate strategic concern in the fifties was Pakistan. Estranged by the trauma of Partition and deadlocked in the territorial dispute over Kashmir, India and Pakistan entered into an arms race that would, in later decades, culminate with both countries' arming themselves with nuclear weapons. In the early years of independence, Pakistan and India competed in the arms race by trying to have a bigger and better-equipped air force than the other country.[23]

India and Pakistan pursued different strategies for modernization and expansion of their air forces, and these strategies were defined by the divergent foreign policies of the two nations. India was nonaligned, while Pakistan signed a mutual assistance agreement with the United States. In return for allowing the United States to use its ports and air bases, Pakistan received American military aid. Through the assistance program, Pakistan received its first American jet fighter, the F-86 Sabre, in 1956.[24]

With the infusion of American military aid, the Pakistan Air Force started to grow faster than the IAF, and India began to worry about being outclassed by its neighbor. The surest way to maintain parity with Pakistan would have been to adopt the same American plane types as the PAF. It was because Pakistan had been given American fighters that the Indian Air Force became preoccupied with fighters from the United States, but it was also because of this very aid that the IAF was never able to acquire any American jet fighters.[25]

Although the United States never sold jet fighters to India, some American noncombat aircraft did enter IAF service, such as the C-119G transport, delivered in 1953. A cargo plane powered by two piston

[22] Vijay Seth, *The Flying Machines: Indian Air Force, 1933 to 1999* (New Delhi: Seth Communications, 2000), 39; Singh, *Indian Aircraft Industry*, 127, 161.

[23] The literature on the India–Pakistan rivalry is immense. See, for example: T.V. Paul, ed., *The India–Pakistan Conflict: An Enduring Rivalry* (New York: Cambridge University Press, 2008); Subrata K. Mitra, "War and Peace in South Asia: A Revisionist View of India–Pakistan Relations," *Contemporary South Asia* 10, no. 3 (November 2001): 361–79.

[24] Robert J. McMahon, *The Cold War on the Periphery: The United States, India, and Pakistan* (New York: Columbia University Press, 1994), 142, 183; John Keay, *Midnight's Descendants: South Asia from Partition to the Present Day* (London: William Collins, 2014), 125, 144; Hussain and Qureshi, *History of the Pakistan Air Force*, 98–101.

[25] "F-104," editorial, *TOI*, August 3, 1961.

Fig. 4.2 An IAF C-119G packet, modified with a booster engine on the top of the fuselage, on display at the IAF Museum in New Delhi, 2012 (Photo by the author)

engines, the C-119G in IAF service provides an example of how local conditions and requirements shaped the aircraft bought from abroad. Originally designed for American climates and conditions where the highest airfield was less than 10,000 feet above sea level, the C-119 suffered from serious performance shortcomings when operating out of the high Himalaya. Hindustan Aircraft, with the assistance of the American firm Steward-Davis, retrofitted twenty-seven IAF C-119s with single turbojet engines mounted on top of the planes' fuselages (Fig. 4.2). This dorsal booster engine provided additional thrust on takeoff, allowing the C-119 to operate out of airfields at elevations as high as 16,800 feet.[26]

Most of the aircraft inducted into IAF service in the fifties were imported, but even from the beginning of the decade the Indian government was taking steps to establish domestic production. In 1950, India signed an agreement with the de Havilland Aircraft Company permitting the manufacture of nearly three hundred Vampire jet fighters under license by HAL. Later, HAL also manufactured Gnat light fighters under a similar agreement with Folland Aircraft. The Gnat served as the IAF's main fighter during the 1960s and into the 1970s. The first twenty-three Gnats inducted by the IAF were manufactured in England and shipped to

[26] Singh, *Aircraft of the IAF*, 93–4.

India complete. The IAF bought its next twenty Gnats as kits, which HAL assembled in Bangalore. All subsequent Gnats purchased by the IAF were built by HAL, from a mixture of domestic and foreign components.[27]

The Indigenous Fighter Dream

In 1956, HAL embarked on its most ambitious project to that point: the design and production of an indigenous jet fighter. The aircraft, which came to be known as the HF-24 Marut, was designed in response to an Air Staff Requirement issued by the Indian government, calling for a Mach-2 jet fighter with a combat radius of 500 mi and an operational ceiling of 60,000 feet. The requirement stipulated that the new jet fighter be produced in India. The Indian Defence Minister, Mahavir Tiagy, asked the Trade Attache at the German embassy in New Delhi, Dr. Taupisch, whether he could recommend an engineer to help develop India's aircraft industry. Taupisch suggested that the Indian government contact Kurt Tank, former chief designer for Focke-Wulf and now an engineer-for-hire.[28]

Kurt Tank would become a key figure in the HF-24 program, leading a team of eighteen German engineers who worked alongside Indian engineers. From 1931 through the end of World War II, Tank had served as the chief designer for the German aeronautical firm Focke-Wulf. His most important design was the FW-190 fighter, of which over 20,000 were built.[29] After the end of the war, Tank was one of many German specialists who carried technical expertise from Germany to the rest of the world—what historian Michael J. Neufeld calls the "Nazi aerospace exodus."

Following the defeat of the Third Reich, German specialists transferred technical expertise from Germany to countries in the eastern and western

[27] In addition to airframe manufacture, HAL built, under license, Orpheus 701 turbojets for the Gnat. HAL would go on to design and produce a highly modified version of the Gnat known as the Ajeet (Unconquered). Between 1962 and 1974, HAL built a total of 192 Gnats for the IAF. Seth, *The Flying Machines*, 39; Singh, *Indian Aircraft Industry*, 146–50; Singh, *Aircraft of the IAF*, 121–22.

[28] Pushpindar S. Chopra, "Harnessing the Storm Spirit," *Air Enthusiast*, May 1973, 218; Heinz Conradis, *Design for Flight*, trans. Kenneth Kettle (London: Macdonald, 1960), 194–95.

[29] "German Engineers Designing India's Jet," *Aviation Week*, December 16, 1957, 79.

blocs, as well as to the nonaligned world. The best-known members of this out-migration of technical talent belonged to a team led by Wernher von Braun, which had developed the V-2 missile for Hitler and went on to design the Saturn V moon rocket for NASA. As Neufeld points out, the spread of German aeronautical expertise was a much larger phenomenon than the relocation of von Braun's team. The technological establishments of the Allied powers all benefited from contact with German engineers. In the case of Tank and others, German engineers also found their way to nonaligned nations.[30]

After World War II, Tank spent nearly twenty years working in nonaligned nations. The defeat of Nazi Germany had put the German aviation industry in limbo, as Germany was no longer building up its military. German engineers who wished to find work in their field—and who had not already been impressed into service by one of the ex-Allied powers—had to seek work abroad. Tank first went to Argentina, where he and his team helped the government of Juan Perón develop its own jet fighter, the Pulqui II. After Perón's ouster in 1955, government support for the jet project evaporated, leaving Tank and his team without solid work.[31]

On invitation from the Indian government, Tank flew from Argentina to India, where he met with Mahavir Tiagy and toured aircraft factories in Bangalore and Bombay. He was suitably impressed by the prospects for aviation in India by the time he returned to Argentina. After prolonged negotiations, Tank left the service of the Argentinian government and relocated to India with his long-time collaborator Ludwig Mittelhüber in August 1956. Some team members in Argentina were lured away to work in the American aerospace industry, but sixteen others ultimately followed Tank and Mittelhüber to India. These eighteen Germans, along with

[30] Michael J. Neufeld, "The Nazi Aerospace Exodus: Toward a Global, Transnational History," *History and Technology* 28 (March 2012): 49–50.

[31] The chief source on Tank's life is Heinz Conradis, *Nerven, Herz und Rechenschieber: Kurt Tank—Flieger, Forscher, Konstrukteur* (Göttingen: Musterschmidt-Verlag, 1955). The book was translated into English by Kenneth Kettle and published as *Design for Flight: The Kurt Tank Story* (London: Macdonald, 1960). Only the later English translation includes an account of Tank's work in India.

Fig. 4.3 HF-24 Marut D-1698, a two-seater trainer version, on display at the HAL Museum in Bangalore in 2015 (Photo by the author)

three Indian senior and twenty-two Indian junior engineers, constituted the HF-24 design team.[32]

The joint Indo-German team set out to design a fighter to meet the Air Staff Requirement. As designed by the team, the HF-24 was a twin-engine jet fighter with a sleek, narrow fuselage and thin, swept wings (Fig. 4.3). The fuselage was area-ruled for superior supersonic performance, which gave it a pinched or coke-bottle appearance. The pilot sat in a Martin-Baker ejection seat beneath a plastic bubble canopy. For armament, the plane carried four 30-mm Aden Mk. II cannons in the nose, as well as fifty air-to-air rockets in the fuselage. In addition, two pylons under each wing permitted the mounting of bombs, drop tanks, or more rockets. At 52 feet in length and weighing 13,700 pounds empty, the Marut was much larger than the mainstay IAF fighter of the 1960s, the Folland Gnat, a 6,600-lb lightweight. Pilots of the Gnat boarded their aircraft by taking a running start and vaulting into the cockpit. On the HF-24, the cockpit rim was above eye-level, and the pilots had to climb up a ladder to board the plane.[33]

The fabrication and testing of HF-24 hardware proceeded in several stages. In standard practice for aircraft projects in the days before

[32] Conradis, *Design for Flight*, 194–95; "India's Supersonic Fighter," *Aeroplane and Astronautics*, October 26, 1961, 548–48a.

[33] Conradis, *Design for Flight*, 201; Singh, *Aircraft of the IAF*, 157–58.

computer-aided design, the plane was first built as a full-scale mockup to verify that the major components fit together properly. The final inspection of the mockup was completed by April 1958, and then the team could proceed with building the first flyable hardware. Before constructing a prototype, HAL made a full-scale wooden glider to test the aerodynamic characteristics of the design. Building a glider was a low-cost substitute for renting time at a wind tunnel overseas. Tank had used the same strategy in Argentina for Pulqui II. Wooden gliders also drew upon Germany's interwar experience, when aeronautical experimentation concentrated on gliders because the Treaty of Versailles placed restrictions on powered flight.[34]

The glider test program began in March 1959 and ran for a year. The size and shape of the glider were the same as the powered fighter. The cockpit's instrument layout was also identical. In the glider, though, the cockpit had room for two pilots seated in tandem. On most of the test flights, the rear seat was occupied by a movie camera, which filmed the instruments as a substitute for more expensive telemetry equipment. The cockpit was too crowded for an ejection seat, so the pilots had to carry their own parachutes. As an added safety measure, the glider also had a spin-recovery parachute, although this was never used in a test.[35]

For the test flights, an Indian Air Force Dakota towed the glider to altitude, then released it so that it could glide steeply back to the airfield. The pilots themselves were IAF officers seconded to HAL. Over the course of about a hundred flights, the pilots tested the stability and low-speed characteristics of the design. Wing Commander Roshan Lal Suri piloted most of the test flights, although Kapil Bhargava flew the last three and became the only person to fly both the unpowered and powered versions of the aircraft. On the last flight, the nose wheel collapsed during landing, damaging the glider and making it unfit to fly again.[36]

After the glider tests, HAL completed the first prototype of the HF-24. Wing Commander Suranjan Das flew the prototype on its first flight out

[34] Chopra, "Harnessing the Storm Spirit," 218; Edward L. Homze, *Arming the Luftwaffe: The Reich Air Ministry and the German Aircraft Industry, 1919–39* (Lincoln: University of Nebraska Press, 1976), 13–15.

[35] "India's Supersonic Fighter," *Aeroplane and Astronautics*, October 26, 1961, 548–48a; Kapil Bhargava, "The HF-24 Glider," *Bharat Rakshak*, updated March 28, 2015, http://www.bharat-rakshak.com/IAF/history/1950s/1224-the-hf-24-glider.html.

[36] Bhargava, "The HF-24 Glider."

of the HAL company airfield, on June 17, 1961. A week later, Das flew the plane on its first official flight, in front of a crowd of 5,000 spectators that included Defence Minister Krishna Menon. In 1962, the IAF ordered eighteen pre-production and sixty-two production HF-24s, the first two of which were delivered in May 1964. It was during this ceremony that the fighter received the name Marut or Storm Spirit. The pre-production Maruts finally entered air force service in 1967, joining the No. 10 Flying Daggers Squadron. By this time, Kurt Tank had moved back to Germany, and the HF-24 team was wholly Indian.[37]

The Air Force and the Superpowers

As the HF-24 prototypes were flying, and even after they had entered squadron service, India was yet far from realizing the dream of equipping the air force with a domestic fighter. HAL was incapable of producing the HF-24 in large numbers, and the planes that did roll out of the factory in Bangalore were underpowered. In attempting to find solutions for these shortcomings, the Indian government looked abroad, to Britain, the United States, and the Soviet Union. Like so much else at the time, India's interactions with Britain and the superpowers were constrained by the politics of the Cold War.[38]

The HF-24's disappointing performance was a result of decisions made by the Indo-German team early in the design process. In an aircraft such as the HF-24, in which the engines are mounted inside the fuselage, engine selection constrains the size and shape of the airframe, so it can be difficult to change this decision once it has been made. The team selected the Orpheus 12, an axial-flow turbojet designed to produce 8,170 pounds of thrust with afterburner. At that time, in the late 1950s, the British firm Bristol Siddeley was still developing the Orpheus 12, and the engine was

[37] "India Enters Supersonic Age: Menon Sees Demonstration by Hindustan Fighter," *TOI*, June 25, 1961; K. Chatterjee, "Hindustan Fighter HF-24 Marut Part 1," Bharat Rakshak, http://www.bharat-rakshak.com/IAF/History/Aircraft/Marut1.html (accessed August 20, 2014); "IAF Receives 2 HAL-Made Supersonic Jets," *TOI*, May 11, 1964.

[38] For the background of South Asia and the Cold War, see McMahon, *Cold War on the Periphery*; Nick Cullather, *The Hungry World: America's Cold War Battle against Poverty in Asia* (Cambridge, MA: Harvard University Press, 2013); Andrew J. Rotter, *Comrades at Odds: The United States and India, 1947–1964* (Ithaca, NY: Cornell University Press, 2000); Bhabani Sen Gupta, *The Fulcrum of Asia: Relations among China, India, Pakistan, and the USSR* (New York: Pegasus, 1970).

not yet ready for commercial production. HAL already had a relationship with Bristol Siddeley, because the Indian firm held a license to produce copies of another engine from the same company, the Orpheus 701, for the Gnat fighter.[39]

As the Orpheus 12 was still not available in 1961 when the HF-24 prototypes were being readied for their first flights, the aircraft used the Orpheus 703, a slightly modified version of the Orpheus 701 with a thrust rating of only 4,850 pounds. HAL planned to install the Orpheus 12 in the production model, but this never happened because Bristol Siddeley ran out of capital and could not complete development of the engine. The British firm offered to continue development if India invested $10 million into the project. India's Ministry of Defence, led by V.K. Krishna Menon, did not want to invest so much hard-earned foreign exchange into an overseas company. The Ministry of Defence sent Bristol Siddeley its regrets and began searching for another engine source.[40]

Ultimately, production aircraft were fitted with the Orpheus 703 and dubbed the HF-24 Mk. 1. The HAL Gas Turbine Division designed an afterburner that raised the thrust of the Orpheus 703 to 6,500 pounds, far short of the performance of the now-canceled Orpheus 12. With two of these engines mounted in its fuselage and running on afterburner, the HF-24 could just reach supersonic speeds. This was too slow for an interceptor, so the IAF used the HF-24 Mk. 1 as a fighter-bomber or ground-attack aircraft. HAL pinned its hopes on the HF-24 Mk. 2, a properly supersonic fighter powered by an engine with higher thrust than the Orpheus 703.[41]

The matter of finding an engine to power the HF-24 Mk. 2 would obsess the project leadership for more than a decade. Even before Bristol Siddeley had definitively canceled the Orpheus 12, HAL had begun considering Soviet engines. One of the engines that held promise was the Klimov VK-7, which had a similar performance to the Orpheus 12. Two VK-7s were shipped from Russia to India for evaluation, but HAL quickly determined that the centrifugal-flow engine was too large to fit

[39] Singh, *Indian Aircraft Industry*, 125; "India Rules Out P.1B Lightning, Widens Search for Strike Fighter," *Aviation Week and Space Technology*, July 23, 1962, 24; Chopra, "Harnessing the Storm Spirit," 218.

[40] "India Rules Out P.1B Lightning"; Chopra, "Harnessing the Storm Spirit," 218; "Current Topics: Jet Engines," *TOI*, October 6, 1961.

[41] "Maruta: India's Hindustan HF-24 Joins the IAF," *Flight International*, July 2, 1964, 17.

Fig. 4.4 An IAF MiG-21, on display at the IAF Museum in New Delhi, 2012 (Photo by the author)

inside the HF-24 fuselage as designed. When the Indians turned their attention to the axial-flow RD-9 engine, which powered the MiG-21, the Russians offered to sell the whole planes instead (Fig. 4.4).[42]

In 1962, Nehru's government signed an agreement with the Soviet Union calling for the purchase of some completed MiG-21s followed by the eventual license manufacture of the single-seat, Mach 2.1 fighters in India. The first six MiG-21s reached Bombay harbor in January 1963, and these aircraft formed the new IAF No. 28 Squadron. HAL began manufacturing MiG-21s at a facility in Nasik, Maharashtra, in 1966. The Soviets allowed the Indian government to pay for the completed aircraft

[42] "India Rules Out P.1B Lightning"; "Current Topics: Jet Engines," *TOI*, October 6, 1961. *Times of India* reported that Bristol Siddeley and the British government refused to share classified specifications of the Orpheus 12 with the Indian government because India had already begun evaluating Soviet turbojets. G.K. Reddy, "Political Factors Creep into Business Deal: U.K. Firm Refuses to Give India Jet Engine Details," *TOI*, October 3, 1961; "No Comment on Jet Engine Issue: U.K. Firm's Refusal," *TOI*, October 4, 1961.

and licensing fees in rupees, which the Soviet government could use only for buying goods in India.[43]

In 1962, the IAF already operated the Soviet Mi-4 helicopter and Antonov An-12 transport, but the MiG-21 was the first Soviet combat aircraft to be adopted in India, as well as the first Russian design of any type to be manufactured there. India would become a faithful customer of military aircraft from Russia. The adoption of the MiG-21 represented the beginning of a shift in the IAF's technological orientation toward the eastern bloc. From a wholly British-equipped and British-trained force at its birth thirty years before, the IAF had begun to adopt Russian influences. In practice, this meant that air force officers and HAL engineers would start traveling to the Soviet Union for training—something that representatives of civilian industries had already been doing for several years. There were also small changes that the IAF would need to make in order to use Russian fighters. Other combat aircraft in IAF service used 30-mm ammunition, but the MiG-21 was equipped with 23-mm cannons, and Indian armaments factories had to retool accordingly. The IAF became even more of a cultural and technological hybrid than the service had been before. Indian pilots, some trained at Cranwell and all English speakers, now flew fighters designed in the Soviet Union.[44]

In the West, India's decision to import combat aircraft from the Soviet Union elicited surprise and dismay. The US government, thinking as usual in binary Cold War terms, saw the MiG deal as India's first step toward joining the eastern bloc. President John F. Kennedy's advisers thought the best response would be to draw India away from the Soviet Union by offering western-bloc arms, even though this strategy ran the risk of angering Ayub Khan, president of Pakistan. The United States and Britain cautiously made an offer to supply India with one squadron of supersonic English Electric Lightning fighters and nine Lockheed C-130 Hercules transports, but only if India back out of the MiG deal. Since this offer was limited in scope and had no provisions for domestic manufacture,

[43] "India Rules Out P.1B Lightning"; Singh, *Aircraft of the IAF*, 142–43; P.C. Lal, Ela Lal (ed.), *My Years with the IAF* (New Delhi: Lancer International, 1986), 123; "Soviet MiG Jets for India Soon: Bid to Counter-Balance Pak Super-Sabrejets," *TOI*, May 6, 1962.

[44] Singh, *Aircraft of the IAF*, 147; Sarwar A. Kashmeri, "MIG-21," letter to the editor, *TOI*, October 19, 1962; HAL, *Annual Report 1964-65*; "Test of Indian Jet Fighter Soon," *Times of London*, March 16, 1961; H.R. Vohra, "HF-24 Project Poses Difficult Problem," *TOI*, August 11, 1964.

Nehru's government found it less appealing than the Soviet offer, and declined.[45]

Another American strategy for luring India away from Soviet combat aircraft was offering assistance for the two open questions of the HF-24: engine procurement and manufacture in quantity. By the time the US government started considering these plans, in 1964, the IAF already had its first MiG-21 squadron, and the MiG factory in Nasik was under construction. The pro-India advisers, who had all been retained by President Lyndon B. Johnson after Kennedy's assassination the year before, hoped that at least India might be persuaded not to close further deals with the USSR. Robert Komer, of the National Security Council, was in favor of a Bristol Siddeley proposal to upgrade the Orpheus 703 to give it a similar performance to the Orpheus 12. "[McGeorge] Bundy and I can't help but feel that Orpheus engine for HF-24 is our secret weapon for side-tracking Soviet MIG and possibly SAM [surface-to-air missile] deals," he wrote to Ambassador Chester Bowles in February 1964. He continued: "You yourself have pointed out how going ahead with HF-24 would also pander to Indian nationalism, while being the course least painful to the Pak[istani]s." If the performance of the HF-24 could be improved with a more powerful engine, everyone would be satisfied.[46]

In July 1964, a delegation of six officers from Wright-Patterson Air Force Base in Dayton, Ohio flew out to Bangalore to inspect the HF-24 program. The officers suggested that a Rolls-Royce engine under development might be compatible with the HF-24 after a few modifications; it was possible that one or two American designs would also work. The western-bloc engines suggested by the American team were still under development, and the Indian government was disinclined to commit to yet another unproven design. Proposals for western-bloc aid for HF-24 production were similarly stillborn. HAL was able to make the HF-24 in small quantities, only about a half-dozen a year. The United States was interested in providing plant machinery that would allow HAL to expand

[45] SNIE 31/32-62, "Possible Reaction of Pakistan to the Provision of Supersonic Fighter Aircraft to India by the US or Other Western Countries," June 6, 1962, *Foreign Relations of the United States*, 1961–63, 19:263 (hereafter cited as *FRUS* with year and volume); Memorandum of Conversation, June 14, 1962, ibid., 19:269; telegram from the embassy in India to the Department of State, June 25, 1962, ibid., 19:292–94; David C. Engerman, *The Price of Aid: The Economic Cold War in India* (Cambridge, MA: Harvard University Press, 2018), 216.

[46] Robert W. Komer to Chester Bowles, February 27, 1964, *FRUS*, 1964–68, 25:46–47; Singh, *Indian Aircraft Industry*, 135.

its production capacity. But Indian government officials made repeated statements that they had no intention of abandoning either the MiG-21 or the HF-24. Since distracting India from the MiG-21 was the reason for aid to the HF-24, the US government dropped any offers of aid.[47]

The United States made one more attempt to draw India away from the MiG-21, by beginning to drop hints again about the possibility of a fighter deal. When Defence Minister Y.B. Chavan visited the United States in May 1964, the Johnson administration suggested that it might be able to supply inexpensive Northrop F-5A Freedom Fighters or Douglas F-6A Skyrays to serve as stopgaps until the HF-24 was ready. Chavan reported in a telegram to Prime Minister Nehru that in Air Vice Marshal Arjan Singh's estimation, the F-5A or F-6A would not "add materially to the efficacy of the IAF." It would be better, Singh counseled, to try holding out for the higher-performance Lockheed F-104G Starfighter.[48]

The Pakistan Air Forcealready had F-104s, and there were differences of opinion within the American defense and diplomatic establishments whether or not India should also receive these aircraft. Defense Secretary Robert McNamara believed that the F-104 was too expensive, too complex, and too politically fraught to sell to India. When approached by Chavan in Washington about the supply of F-104s to India, McNamara said that the current assistance program could not be expanded to accommodate the desired fighter.[49]

Disagreeing with McNamara, Robert Komer felt that the United States should offer F-104s to India, along with F-6As and assistance for the HF-24 program. He wrote to McGeorge Bundy after Chavan's visit that, although "Bob McNamara is dead right that F-104s are too rich for India's blood," military aid was a matter of prestige for the United States. In reference to several recent failed aid ventures, Komer commented that "Soviets are picking up too many options we let drop—Bokaro, VOA

[47] "Improving HF-24: Six US Experts in India," *TOI*, July 16, 1964; "Airframe Needs Changes: U.S. Team's Views on HF-24," *TOI*, August 7, 1964; "Both MIG & HF-24 Planes to Be Built," *TOI*, September 22, 1964; "American Aid 'Uncertain'," *TOI*, April 27, 1965; Jacques Nevard, "U.S. May Aid India on Improving Jets," *NYT*, July 16, 1964; Evert Clark, "U.S. Offers India Aid on Air Force," *NYT*, August 22, 1964.

[48] Y.B. Chavan, telegram to Jawaharlal Nehru, May 22, 1964, S. No. 18, Ambassador to USA subject files, papers of B.K. Nehru, NMML.

[49] Y.B. Chavan, telegram to Jawaharlal Nehru, May 22, 1964, S. No. 18, Ambassador to USA subject files, papers of B.K. Nehru, NMML.

transmitter, supersonics, etc." Supplying an F-104 squadron to India was "about [the] only big gesture we have currently available."[50]

Ultimately, it was McNamara's thinking that would prevail. By the beginning of 1965, India's defense establishment realized that it was not likely to receive F-104s any time soon, and thus decided to change tack. In a meeting with Ambassador Bowles in late January, Defence Minister Chavan reported that his government was interested in the offer of F-5As after all. These aircraft could take over the ground support role of the aging Vampires, Toofanis, and Mystères, which were becoming increasingly expensive to maintain.[51]

Discussion about a possible arms sale dragged on for several months. In March, Ambassador B.K. Nehru pressed McNamara to expedite the supply of aircraft. The American Defense Secretary replied that his government was having trouble locating surplus equipment, but he would try to speed up the process. Within weeks, though, the military assistance program had been derailed by the outbreak of fighting between India and Pakistan in the Rann of Kutch, a prelude to the 1965 Indo-Pakistani War. On May 8, Secretary of State Dean Rusk advised Bowles that the US government had decided not to fulfill the Indian request for fighters.[52]

India, Egypt, and Indigenous Jet Fighters

The Indian government continued to search for an engine to power the HF-24 Mk. 2. Having explored options in both the western and eastern power blocs, India turned to the nonaligned world. In Egypt, India found a potential technological donor and collaborator. The United Arab Republic (as Egypt was still officially known despite the failure of a short-lived union with Syria in 1961) was led by Gamal Abdel Nasser, a

[50] Memo, Robert Komer to McGeorge Bundy, May 27, 1964, *FRUS*, 1964–68, 25:109–10.

[51] "Aide Memoire—Acquisition of F5A from USA," n.d. [January 28, 1965], S. No. 18, Ambassador to USA subject files, papers of B.K. Nehru, NMML; R.J. Rebello to Charles E. Johnson, February 15, 1965, ibid.; P.V.R. Rao to B.K. Nehru, March 6, 1965, ibid.

[52] B.K. Nehru, telegram to P.V.R. Rao, March 19, 1965, S. No. 18, Ambassador to USA subject files, papers of B.K. Nehru, NMML; telegram, Chester Bowles to State Department, January 29, 1965, *FRUS*, 1964–68, 25:184–86; telegram, McGeorge Bundy to Chester Bowles, April 28, 1965, ibid., 25:240–42.

modernizing elite who shared with Nehru an interest in industrialization and economic self-reliance.[53]

Nasser's Egypt, like Nehru's India, had also tapped into the Nazi aerospace exodus. In the fifties, the Egyptian government recruited rocketry experts to help the country develop its armaments industry, but these engineers returned to Germany after being targeted by Israeli hit squads. Later, in the early sixties, UAR acquired its own indigenous jet fighter program, the HA-300. Led by Willy Messerschmitt, the HA-300 program had begun in Franco's Spain, where Messerschmitt had migrated after spending a brief period in India advising HAL. Messerschmitt and his team of expatriate engineers designed the HA-300 (the initials standing for Hispano Aviación) around the Orpheus 12 engine, the same stillborn design for which the HF-24 had also been designed.[54]

The HA-300 test program began with flights of a full-scale glider, towed into the air by a CASA C-2.111 (the Spanish-built version of the German Heinkel He-111 bomber). The glider proved unstable, necessitating a major redesign, but by this time the Spanish government had lost interest. Nasser's government happened to be looking for a new project, so it bought the HA-300 off of the Spanish. The Egyptian project kept the name it had had in Spain, but now the initials stood for "Helwan Aircraft," after the Cairo suburb where the project was located.[55]

Since the Orpheus 12 was now a definite failure, the Egyptian government expanded the scope of the HA-300 project to include development of a jet engine, the E-300. A domestic design, the E-300 was built primarily from imported materials and parts. Ferdinand Brandner, an Austrian engineer who had previously been impressed into designing turboprop engines for the Soviet Union, oversaw development of the E-300 in Factory 135 at the Helwan airfield. Hassan Kamil, an Egyptian businessman residing in Switzerland, established an agency for coordinating the recruitment of European talent and the import of tools

[53] For Egypt's Techno-Statism, see Timothy Mitchell, *Rule of Experts: Egypt, Techno-Politics, Modernity* (Berkeley: University of California Press, 2002).

[54] Neufeld, "Nazi Aerospace Exodus," 55; A.C.N. Nambiar to S.A. Venkataraman, March 18, 1949, file no. 20(20)-EurII/49, Ministry of External Affairs, NAI.

[55] Kapil Bhargava, "HA-300: Egypt's Cosmopolitan Status Symbol," *Air Enthusiast*, November 1979–February 1980, 2.

and parts. About 470 technical specialists from Spain and the German-speaking countries joined the HA-300 and E-300 projects at Helwan Aircraft.[56]

Watching these developments from afar, the Indian government sensed an opportunity for technical collaboration. Egypt had a jet engine but India did not; for its part, India had a working aircraft prototype, experienced test pilots, and generally deeper aeronautical expertise than Egypt. The Egyptians had been unable to find a qualified European test pilot to fly the HA-300 prototype, so the Indian Air Force seconded Squadron Leader Kapil Bhargava to Helwan Aircraft in June 1963. Later, several other Indian pilots and navigators joined Bhargava in Cairo. Starting with taxiing tests in February 1964 and the first flight the following month, Bhargava tested two prototypes of the HA-300. Years later, he would recall that the HA-300 was "very pleasant to fly."[57]

The Indo-UAR collaboration entailed transfer of expertise and development of human capital, a form of technology circulation. In addition to testing the HA-300 prototypes, Bhargava trained two Egyptian pilots for test flying. Only one Egyptian pilot had ever enrolled in the Empire Test Pilots' School at Farnborough, England, but he had been killed in a crash during training. After 1956, when Egypt provoked the ire of France and Britain by seizing the Suez Canal, Empire and other schools in western Europe became off-limits to Egyptian pilots. Bhargava, a graduate of the Empire Test Pilots' School, trained two Egyptian Air Force pilots, Major Zohair Shalaby and Major Sobhy el Tawil, in UAR and India. The Egyptian pilots performed some of the test flying of the HA-300 and a parallel trainer project, the HA-200.[58]

The HA-300 airframe was ready for flight tests years before the E-300 engine, so the prototype fighters were powered by the same engine as the HF-24 Mk. 1, the Orpheus 703. When the E-300 was ready for testing, Helwan Aircraft ran it in static test cells and in flight under the wing of an

[56] Bhargava, "HA-300: Egypt's Cosmopolitan Status Symbol," 2; Neufeld, "Nazi Aerospace Exodus," 56; "High Speeds Now Made Possible: HF-24 Aircraft," *TOI*, October 2, 1964.

[57] Bhargava, "HA-300: Egypt's Cosmopolitan Status Symbol," 3–5; Bhargava, "Messerschmitt's HA-300 and Its Indian Connection"; "IAF Engineers in Cairo," *TOI*, May 24, 1966.

[58] Bhargava, "HA-300: Egypt's Cosmopolitan Status Symbol," 5.

Antonov An-12 transport. The latter tests were not particularly informative, because they represented only a limited portion of the engine's total performance envelope. As part of an expanded collaboration agreement signed in September 1964, India sent a modified HF-24 to Egypt to serve as an E-300 testbed. The aircraft, known as IBX, had an Orpheus engine on its port side and an E-300 on starboard. After more than a hundred flights, mostly performed by IAF Wing Commander I.M. Chopra, the joint Indo-Egyptian team working on the project declared the E-300 safe for the single-engine HA-300.[59]

The two HA-300 prototypes survived an Israeli air raid on Helwan during the Six-Day War in June 1967, but within two years the indigenous jet program was dead. By October 1968, Air Marshal P.C. Lal, the Managing Director of HAL, had begun backpedaling, stating that India was not committed to buy the engine being developed for the Egyptian fighter. If a better engine were to appear from another source, India would buy that one and use it in the HF-24 Mk. 2. By the following year, the Egyptian government had also lost interest, finally shutting down the HA-300 project in May. Test pilot Bhargava was there to the bitter end. Helwan Aircraft officials received authorization to perform a few more tests after the official end of the project. A third HA-300 prototype had been built and equipped with an E-300, but not yet flight-tested. The prototype was taxied under power, but the ground crew found that the oil system had been contaminated by fuel, so the plane was never taken up for a test flight.[60]

By the time of the failure of the Indo-UAR collaboration, it was already clear that the HF-24 was not going to become the mainstay of the air force any time soon. The IAF went shopping for foreign fighter-bombers to serve as intermediate ground-support aircraft until the HF-24 was ready. After negotiations for the US-built F-5A fell through, the Indian government began to consider the Sukhoi Su-7, a Soviet design. Since this plane seemed to offer a good tradeoff of cost and performance, the government placed an order for a hundred complete aircraft, the first

[59] Bhargava, "HA-300: Egypt's Cosmopolitan Status Symbol," 7; "IAF Engineers in Cairo," *TOI*, May 24, 1966; "Tests on HF-24 in UAR Continue," *TOI*, April 2, 1968.

[60] Bhargava, "HA-300: Egypt's Cosmopolitan Status Symbol," 8; "No Commitment with UAR: HF-24 Engine," *TOI*, October 4, 1968.

of which arrived in crates at Bombay harbor in March 1968. With the purchase of such a large number of foreign aircraft, India's quest for an indigenously produced air force remained elusive.[61]

Conclusion

In the preface to his 2011 study *Indian Aircraft Industry*, retired IAF officer Air Commodore Jasjit Singh describes the strategy followed by Nehru's government for achieving self-reliance of military equipment. "Aircraft started to be acquired with a three-strand strategy," he writes, "(i) by direct purchase; (ii) through licensed production; and (iii) by initiating indigenous design and development." Singh correctly notes that there was never an official articulation of this strategy by the government. In fact, it was hardly a coherent strategy, as it is visible only in hindsight. Nehru may have envisioned such a procession from import to domestic production to indigenous design, but the IAF leadership, with its preoccupation with buying the most sophisticated fighters to counter Pakistan, was not committed to such a gradualist approach.[62]

Why did India's first indigenous jet fighter program fail to produce a reliable import substitute? It was not from lack of technical ability or political commitment. Rather, it failed because of economic factors, both those specific to this project and others inherent to an import-substitution program of this magnitude. In the specific case, the HF-24 was designed around an unproven engine that never went into production. The replacement engine, the Orpheus 703, was not powerful enough for the plane to fill its original role of interceptor, so the plane had to be repurposed as a fighter-bomber. As Michael Neufeld notes, engine supply was the "Achilles heel" of indigenous jet programs. Designing an airframe was easy compared with building an engine.[63]

More broadly speaking, Indian industry did not have the capacity to produce sufficient quantities of jet fighters. India had HAL, but HAL could not work alone. As a prime contractor, it needed to work in concert with a network of subcontractors and parts suppliers. HAL could build an airplane and an engine—albeit an underpowered one—but it could

[61] Singh, *Aircraft of the IAF*, 164.

[62] Singh, *Indian Aircraft Industry*, xii.

[63] Neufeld, "The Nazi Aerospace Exodus," 58.

not make ejection seats, machine guns, landing gear struts, instruments and gauges, or countless other small but essential parts. Fifty percent of the components in the early HF-24s were imported from Britain, including hydraulic systems, wheels, tires, and brakes. Over time, as parts suppliers were established in India, the import component dropped below 50 percent, but it never even got close to zero.[64]

The history of the Indian Air Force is, and always has been, more than the story of indigenous jet fighter programs. Building an Indian jet was an idealistic venture, but the reality is that the modernization of the air force in the early independence period—or any period—relied more on importation than indigenous production. In addition to aircraft, the IAF also imported search radar, ground support equipment, and ideas and practices.

The Indo-Pakistan arms race was a major motivation for Indian militarization in the early decades of independence, but it was not the only one. It would, in fact, be war with China in 1962 that would compel India to rethink its strategy and the relationship between defense, autarky, and nonalignment.

References

1. Bassett, Ross. *The Technological Indian*. Cambridge, MA: Harvard University Press, 2016.
2. Bhargava, Kapil. "HA-300: Egypt's Cosmopolitan Status Symbol." *Air Enthusiast*, November 1979–February 1980, 1–8.
3. Bhargava, Kapil. "India's Reclaimed Bombers." Bharat Rakshak. http://www.bharat-rakshak.com/IAF/aircraft/past/927-b24.html.
4. Bhargava, Kapil. "The HF-24 Glider." Bharat Rakshak. http://www.bharatrakshak.com/IAF/history/1950s/1224-the-hf-24-glider.html.
5. Chatterjee, K. "Hindustan Fighter HF-24 Marut Part 1." Bharat Rakshak. http://www.bharatrakshak.com/IAF/History/Aircraft/Marut1.html.
6. Conradis, Heinz, and Kenneth Kettle, trans. *Design for Flight: The Kurt Tank Story*. London: Macdonald, 1960.
7. Cullather, Nick. "Damming Afghanistan: Modernization in a Buffer State." *Journal of American History* 89, no. 2 (2002): 512–37.

64 "India Enters Supersonic Age: Menon Sees Demonstration by Hindustan Fighter," *TOI*, June 25, 1961; "India's Supersonic Fighter," *Aeroplane and Astronautics*, October 26, 1961, 548–48a.

8. Devereux, David R. "The Sino-Indian War of 1962 in Anglo-American Relations." *Journal of Contemporary History* 44, no. 1 (2009): 71–87.
9. Edgerton, David. *The Shock of the Old: Technology and Global History Since 1900*. Oxford: Oxford University Press, 2007.
10. Engerman, David C. *The Price of Aid: The Economic Cold War in India*. Cambridge, MA: Harvard University Press, 2018.
11. Gray, Peter W. "The Myths of Air Control and the Realities of Imperial Policing." *Aerospace Power Journal* 15, no. 3 (fall 2001): 21–31.
12. Gunston, Bill. *Mikoyan MiG-21*. London: Osprey, 1986.
13. Homze, Edward L. *Arming the Luftwaffe: The Reich Air Ministry and the German Aircraft Industry, 1919–39*. Lincoln: University of Nebraska Press, 1976.
14. Hussain, Syed Shabbir, and M. Tariq Qureshi. *History of the Pakistan Air Force, 1947–1982*. Karachi: PAF Press, 1982.
15. Keay, John. *Midnight's Descendants: South Asia from Partition to the Present Day*. London: William Collins, 2014.
16. Khan, Yasmin. *The Great Partition: The Making of India and Pakistan*. 2007. Reprint New Delhi: Penguin, 2013.
17. Leake, Elisabeth Mariko. "British India Versus the British Empire: The Indian Army and an Impasse in Imperial Defence, Circa 1919–39." *Modern Asian Studies* 48, no. 1 (2014): 1–29.
18. Meilinger, Phillip. "Trenchard and 'Morale Bombing': The Evolution of Royal Air Force Doctrine before World War II." *Journal of Military History* 60 (1996): 243–70.
19. Moraes, Frank, ed. *The Indian and Pakistan Year Book and Who's Who 1951*. Bombay: Times of India, 1951.
20. Moraes, Frank. *Jawaharlal Nehru: A Biography*. New York: Macmillan, 1956.
21. Moy, Timothy. "Transforming Technology in the Army Air Corps, 1920–1940: Technology, Politics, and Culture for Strategic Bombing." In *The Airplane in American Culture*, edited by Dominick A. Pisano, 299–332. Ann Arbor: University of Michigan Press, 2003.
22. Neufeld, Michael J. "The Nazi Aerospace Exodus: Toward a Global, Transnational History." *History and Technology* 28 (March 2012): 49–67.
23. Omissi, David E. *Air Power and Colonial Control: The Royal Air Force, 1919–1939*. Manchester: Manchester University Press, 1990.
24. Paris, Michael. "Air Power and Imperial Defence 1880–1919." *Journal of Contemporary History* 24, no. 2 (1989): 209–25.
25. Seth, Vijay. *The Flying Machines: Indian Air Force, 1933 to 1999*. New Delhi: Seth Communications, 2000.
26. Singh, Jasjit. *Indian Aircraft Industry*. New Delhi: KW Publishers, 2011.

27. Singh, Jasjit. *The Icon: Marshal of the Indian Air Force Arjan Singh, DFC—An Authorised Biography*. New Delhi: KW Publishers, 2009.
28. Singh, Pushpindar. *Aircraft of the Indian Air Force, 1933–73*. New Delhi: English Book Store, 1974.
29. Singh, Pushpindar. *History of Aviation in India: Spanning the Century of Flight*. New Delhi: Society for Aerospace Studies, 2003.

CHAPTER 5

India's Air Defenses and the Sino-Indian War: Exercise Shiksha, November 1963

At dawn on October 20, 1962, the Chinese People's Liberation Army launched a two-front offensive in the Himalayan heights along the disputed border between China and India. In the Northeast Frontier Agency (now Arunachal Pradesh), PLA troops crossed two 16,000-foot passes and routed the Indian border sentries stationed there. In Ladakh to the northwest, the Chinese troops quickly overpowered the Indian *jawans* (soldiers) defending army posts.[1]

Coming after three and a half years of escalating tensions, in which Indian provocations had played no small part, the Chinese attack was hardly a surprise to New Delhi.[2] But as the PLA moved deeper into the

[1] Srinath Raghavan, "A Bad Knock: The War with China, 1962," in *A Military History of India and South Asia, from the East India Company to the Nuclear Era*, ed. Daniel Marston and Chandar S. Sundaram (Westport, CT: Praeger Security International, 2007), 157–74.

[2] Xuecheng Liu notes in *The Sino-Indian Border Dispute and Sino-Indian Relations* (Lanham, MD: University Press of America, 1994), 10, that early historiography of the conflict showed sympathy to India by portraying China as an aggressor. This changed with Neville Maxwell's *India's China War* (London: Cape, 1970); and Karunakar Gupta's *Hidden Story of the Sino-Indian Frontier* (Calcutta: Minerva Associates, 1974), two books highlighting the political and strategic missteps made by Nehru and his government. Maxwell based his conclusions heavily on the Henderson Brooks-Bhagat Report, a classified Indian Army document produced in 1963. The report criticized the Indian

W. A. T. Logan, *A Technological History of Cold-War India, 1947–1969*, Palgrave Studies in the History of Science and Technology,
https://doi.org/10.1007/978-3-030-78767-7_5

high valleys along both fronts, the halls of Sansad Bhawan (Parliament House) rang out with recriminations as both Congress and the opposition searched for scapegoats to blame for the failings of the Indian military. Nehru dismissed his Defence Minister, V.K. Krishna Menon, who had served the Congress government since independence.

By November 19, the Indian government was in dire straits. In NEFA, the Chinese had taken the town of Tawang and seemed on the verge of breaking out of the mountains and into the indefensible flatlands of the Brahmaputra Valley. Once that happened, nothing could stop them from taking Tezpur on the north bank of the Brahmaputra. Civilians fled from Tezpur as policemen burned government records to keep them out of the hands of the Chinese.[3] Watching the situation from New Delhi, Nehru felt trapped between two undesirable options: capitulating to the Chinese or repudiating nonalignment by requesting military intervention from a foreign power.

Reluctantly choosing the latter option, Nehru dispatched two diplomatic cables to President John F. Kennedy in Washington. In the first, he painted a grim picture of the strategic situation, with the Chinese holding most of NEFA in the east and threatening to overrun Chushul in the west, the last defensible point before Leh, the headquarters of Ladakh. "Events have moved very fast," he wrote, "and we are facing a grim situation in our struggle for survival and in defending all that India stands for against an unscrupulous and powerful aggressor." The Prime Minister stated cryptically that jet fighters and transport planes were "absolutely necessary to stem the tide of Chinese aggression," but he stopped short of requesting that the United States provide these necessities.[4]

Within a few hours, Nehru felt that he could no longer afford to be indirect. In a second telegram, he declared that the situation had become "really desperate." With NEFA all but overrun by the Chinese, the oil

Army's management of the border dispute. In March 2014, Maxwell leaked the report online. The unauthorized release of the report has prompted a rethinking of the Sino-Indian War. See Sandeep Unnithan, "Henderson Brooks Report Lists the Guilty Men of 1962," *India Today*, March 18, 2014, http://indiatoday.intoday.in/story/the-guilty-men-of-1962-india-china-war-jawahar-lal-nehru-krishna-menon/1/350080.html.

[3] "Evacuees Reach Gauhati," *Assam Tribune*, November 22, 1962; "Fight on Border Halted, Nehru Tells Parliament; China Reports Cease-Fire," *New York Times* (*NYT*), November 22, 1962.

[4] Jawaharlal Nehru to John F. Kennedy, November 19, 1962 (first telegram), Nehru correspondence, National Security Files, Presidential Papers, JFK Library.

fields of Assam were directly threatened. In Ladakh, the Chinese had begun shelling the airfield at Chushul. Across the border in Tibet, the Chinese air force was more active than usual. To respond to the heightened threat, Nehru asked Kennedy to send radar stations and twelve squadrons of all-weather fighters to protect Indian forces and cities. Since Indians did not yet have the training to operate this equipment, the radar and fighters would have to be crewed by American personnel. In other words, Nehru was not only requesting military aid, but direct intervention. "We are confident," the Prime Minister wrote in conclusion, "that your great country will in this hour of our trial help us in our fight for survival and for the survival of freedom and independence in this sub-Continent as well as the rest of Asia."[5]

It was an astonishing proposal, born of the trauma of defeat. Had Kennedy agreed to Nehru's request, the consequences for Indian domestic policy, nonalignment, and American foreign policy would doubtless have been far-reaching. But Kennedy never even replied to Nehru's messages before the end of the war. The day after Nehru sent his desperate telegrams, the Chinese made a unilateral cease-fire offer and proceeded to withdraw to the positions they had held before October 20. Although India regained most of its lost territory, the war made a deep impression on the Indian government, military, and general public. As Sarvepalli Gopal wrote, "No one who lived in India through the winter months of 1962 can forget the deep humiliation felt by all Indians, irrespective of party."[6]

Even though Kennedy did not send air cover for Nehru—and may not have even if he had had the chance—he did respond to Nehru's requests for military aid after the fighting stopped. In the year after the conflict, Nehru's request passed through diplomatic meetings and military conferences, ultimately becoming a joint Indo-US-UK military exercise codenamed Shiksha, after a Sanskrit word for education. In November 1963, fighters from the American and British air forces conducted practice intercepts alongside Indian Air Force fighters, using IAF and Royal Australian Air Force bombers as mock enemy targets. The US Air Force

[5] Jawaharlal Nehru to John F. Kennedy, November 19, 1962 (second telegram), Nehru correspondence, National Security Files, Presidential Papers, JFK Library.

[6] Sarvepalli Gopal, *Jawaharlal Nehru: A Biography* (Cambridge, MA: Harvard University Press, 1984), 3:232; Robert J. McMahon, *The Cold War on the Periphery: The United States, India, and Pakistan* (New York: Columbia University Press, 1994), 292–93.

supplied temporary radar sets for use in the exercise, with plans for permanent sets to replace them later.

The Indian government presented Exercise Shiksha as an opportunity for the IAF to get training in up-to-date interception techniques and the use of American radar. This did take place in Exercise Shiksha; but the exercise was also an opportunity for western forces to get accustomed to operating in India should they ever be called on for help in the future.

In the last chapter, we saw how India's quest for modernization and autarky ran up against practical economic considerations. In this chapter, Exercise Shiksha represents another limit: strategic realities. The Sino-Indian War discredited Nehru's vision that India could lead the nonaligned world by moral force rather than military might. The war led to vastly increased military expenditures and a greater dependence on imports. Autarky was all well and good, but the military could not sit idly by and wait for industry to catch up as China—and Pakistan—threatened.

India and China Go to War

To understand the roots of the 1962 Sino-Indian War, we could look all the way back to July 1914, when the British Raj hosted a trilateral border conference at the imperial summer capital of Simla. In the Simla Conference, British India, Tibet, and China agreed on an international border running along the crest of the eastern Himalaya, on an alignment known as the McMahon Line. But we really need only look back to the 1950s, when tensions began to build over contested territory. Although both India and China used the Simla Conference to their own political ends—India to prove its right to NEFA, China to claim that it had been bullied into making unfair territorial concessions—what matters is not so much the substance of the colonial legacy as the way this legacy was used in the fifties. After all, none of the three parties of the Simla Conference still existed in the same form as in 1914. British India was now independent India; republican China was the communist People's Republic of China; and Tibet as a semi-sovereign state had ceased to exist.[7]

[7] Shanti Prasad Varma, *Struggle for the Himalayas: A Study in Sino-Indian Relations* (Jullundur, India: University Publishers, 1965), 15; Xuecheng Liu, *The Sino-Indian Border Dispute and Sino-Indian Relations* (Lanham, MD: University Press of America, 1994), 26.

There was nothing inevitable about the Sino-Indian War. In the fifties, China and India actually had much common ground. The two largest countries in Asia, both had recently emerged from centuries of foreign domination and were attempting to rebuild and modernize their societies. Although Prime Minister Nehru and Chairman Mao had radically different political philosophies, both believed that their countries could best develop under socialistic command economies and centralized planning. Even though Nehru was dubious about the legality of China's annexation of Tibet in 1950, in 1954 he signed a five-point agreement that, among other things, recognized China's right to Tibet.[8]

Chinese and Indian territorial claims overlapped in several places along the border. In the west, between Pakistan and Nepal, India and China both claimed Aksai Chin, a barren and largely uninhabitable plateau (Fig. 5.1). Never politically a part of India, Aksai China had been claimed by the British in 1911, when the newly minted Chinese republic was too weak to protest. To the southeast, a small area of the border of Uttar Pradesh (now Uttarakhand) was also claimed by both countries. On the other end of the frontier, south of the McMahon Line, China claimed the southern slopes of the Himalaya, which were administered by the Indian government as the Northeast Frontier Agency (Fig. 5.2).[9]

As late as 1950, the Indian government published maps designating the Himalayan borders of Ladakh and Uttar Pradesh as "undefined." But then the Indian government began to double down on its territorial claims. In February 1951, an Indian political officer moved into Tawang, displacing the Tibetan administrators in the city. The Indian government followed up on its assertion to NEFA by building roads and setting up border checkpoints. The government also published new maps with clearly demarcated borders, and tried to retract the earlier maps.[10]

Meanwhile, China was asserting its own claims, by publishing maps that showed large parts of India in Chinese territory. The government also built a road from Sinkiang (Xingjiang) to Tibet, passing through Aksai Chin. Indian officials had known about the road during its construction,

[8] McMahon, *Cold War on the Periphery*, 216; Srinath Raghavan, *War and Peace in Modern India* (New Delhi: Permanent Black, 2010), 240; Frank Moraes, *Jawaharlal Nehru: A Biography* (New York: Macmillan, 1956), 454.

[9] Raghavan, *War and Peace in Modern India*, 229.

[10] Raghavan, *War and Peace in Modern India*, 228, 235, 240, 242.

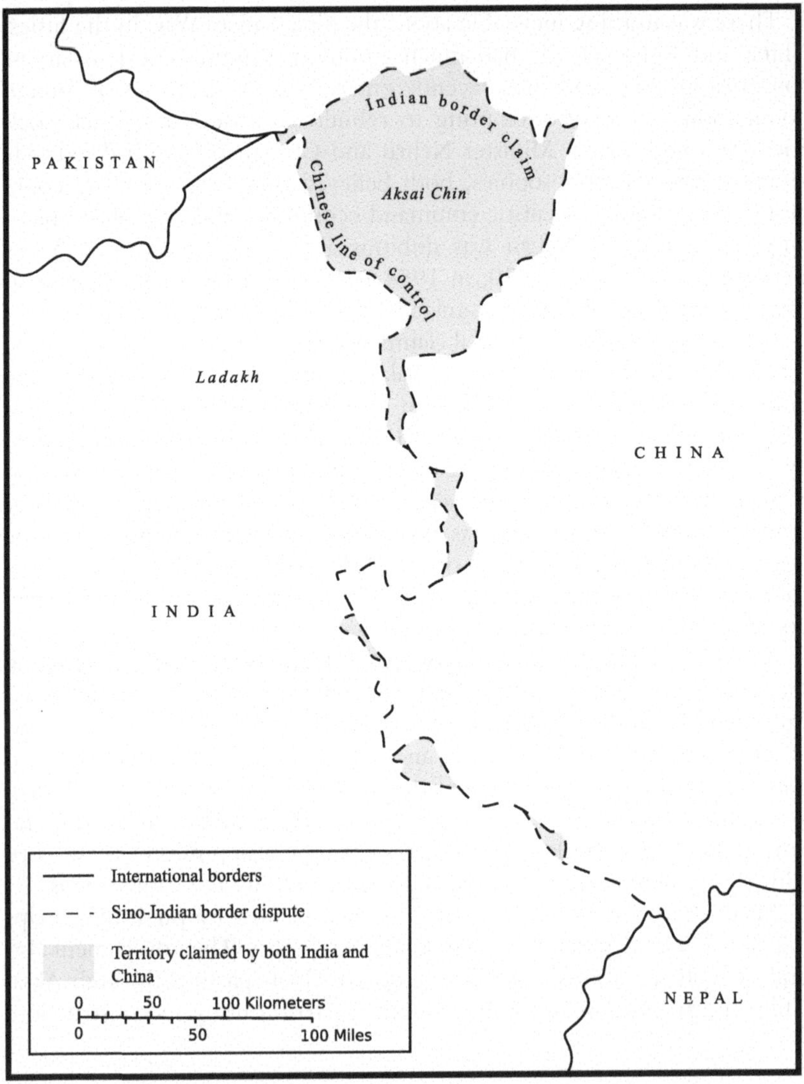

Fig. 5.1 The western portion of the China–India border, showing the contested area (Map drawn by author, based on map at http://www.lib.utexas.edu/maps/middle/east/and/asia/china/indiaw/border/88.jpg, accessed March 2, 2012)

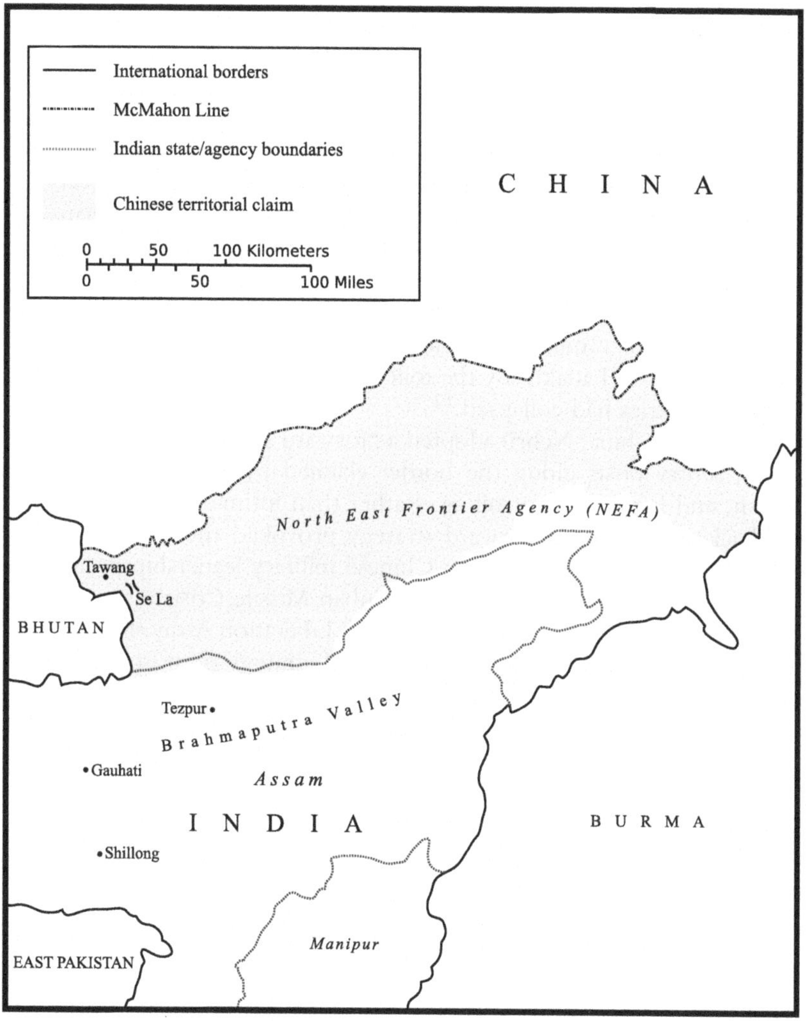

Fig. 5.2 The eastern section of the China–India border, showing the contested area (Map drawn by author, based on map at http://www.lib.utexas.edu/maps/middle/east/and/asia/china/india/e/border/88.jpg, accessed March 2, 2012)

but it was not until a press release announced its completion in September 1957 that they realized it passed through Aksai Chin.[11]

Up to this point, the border dispute had been a bloodless quarrel played out by road-building and map-printing. Then in 1959, Sino-Indian tensions boiled over into armed clashes. In March, the 23-year-old Dalai Lama of Tibet fled his homeland and received asylum from the Indian government. China was incensed. Within months, Indian and Chinese border troops had begun skirmishing on the long frontiers between the two countries. On August 25, Indian troops ambushed Chinese guards at Migyitun, on the Tibetan side of the McMahon Line. Two months later, Indian police forces were killed in a Chinese ambush at Kongka Pass in Ladakh. By the following summer, diplomacy between the two countries had collapsed.[12]

To counter China, Nehru adopted a "forward strategy," which entailed placing sentry posts along the border claimed by India in NEFA and Ladakh, and increasing patrolling. Rather than intimidating the Chinese into backing down, the forward strategy provoked them. On October 6, 1962, Chairman Mao and the Chinese military leadership decided to attack India. Two weeks later, as the Cuban Missile Crisis distracted the attention of the major powers, the People's Liberation Army stormed into the territory claimed by India. The forward strategy outposts in Ladakh were the first to fall to the Chinese.[13]

From their positions on the Tibetan Plateau, the Chinese had a distinct geographical advantage. The southern slope of the Himalaya is pitched steeply upward, so the Indians trying to defend this territory had to struggle up steep roads that were frequently wiped out by mudslides. Trucks broke down or got stuck and had to be dragged out inch by inch by groups of *jawans*. Some parts of the front on the Indian side could be accessed only by mule train. The Chinese, for their part, had good roads crossing the comparatively gentle terrain of the Tibetan Plateau.[14]

[11] Raghavan, *War and Peace in Modern India*, 242, 245.

[12] Raghavan, *War and Peace in Modern India*, 251, 257, 266; Larry M. Wortzel, "Concentrating Forces and Audacious Actions: PLA Lessons from the Sino-Indian War," in *The Lessons of History: The Chinese People's Liberation Army at 75*, ed. Laurie Burkitt, Andrew Scobell, and Larry M. Burkitt (Carlisle, PA: Strategic Studies Institute, 2003), 331–32.

[13] Raghavan, *War and Peace in Modern India*, 276, 299, 304.

[14] "Up High, Up Front, in an Inscrutable War," *Life*, November 30, 1962, 26–33.

On the homefront—in the cities, towns, and villages of India—the public expressed anger and patriotic resolve. Although politicians' speeches and the oversized headlines of newspapers exaggerated the people's unity and patriotism, many of these reactions were no doubt genuine. Chinese Premier Zhou Enlai was burned in effigy and angry students marched on Delhi carrying banners demanding "CHINESE OUT." Young men raced to recruiting offices to enlist, young women joined the National Cadet Corps, and matrons donated their gold jewelry to the war effort. Whether directly or indirectly, the war affected almost everyone in India. In Gauhati, not far from the fighting in NEFA, the festival of Diwali was celebrated without the usual light displays or fireworks.[15]

In New Delhi, Nehru had been awoken from his dreams of an untouchable nonaligned India by a gust of Chinese wind. Shortly after the attack began, the Prime Minister sent letters to the leaders of friendly nations, asking for "the sympathy and support of all countries." Half a world away in Washington, the Kennedy administration watched the events in India closely, even as US–Soviet tensions mounted in the Cuban Missile Crisis. Nehru's letter did not explicitly mention military aid, but Kennedy read between the lines and offered to supply India with small arms, ammunition, and other equipment for mountain warfare. This was the beginning of the American Military Assistance Program (MAP) to India. Nehru gladly accepted Kennedy's offer. US Air Force Boeing C-135 jets flew the aid shipments to Delhi and Calcutta, and Lockheed C-130 Hercules transports shuttled the supplies up to isolated frontier areas. Between November 1962 and June 1963, the USAF flew more than 3,000 missions in India, supplementing the Soviet-built Antonov An-12s and American-built Fairchild C-119Gs used by the IAF (Fig. 5.3).[16]

[15] "The War That Shook the Giant," *Life*, November 16, 1962, 46–49; "A Vast Disillusion Shatters India," *Life*, November 9, 1962, 53–56; "The Nation Responds," *Sainik Samachar*, November 11, 1962; "Dewali Celebrated in Subdued Way," *Assam Tribune*, October 28, 1962; "'Chou En-lai' Burnt 50,000 Times," *Assam Tribune*, November 4, 1962.

[16] Jawaharlal Nehru, *Nehru Writes to Heads of States* (New Delhi: Ministry of External Affairs, n.d. [1962]), 4, in "Foreign Relations of India, Part 1," *South Asia Ephemera Collection* (New Delhi: Library of Congress Office, 1997); McMahon, *Cold War on the Periphery*, 287; David R. Devereux, "The Sino-Indian War of 1962 in Anglo-American Relations," *Journal of Contemporary History* 44, no. 1 (2009): 71–87; History of the 2d Weather Wing, July–December 1963, Call # K360.2-2, IRIS # N3726, in the USAF Collection, Air Force Historical Research Administration, Maxwell AFB, AL.

Fig. 5.3 A visibly haggard Nehru conferring with US Ambassador Galbraith at the time of the border war with China (*Source* JFK Library, JFKPOF-118a-004)

On November 20, 1962, the Chinese government made its surprise cease-fire offer and began to withdraw. Although the Chinese government was characteristically close-mouthed about its motivations, it appears that China never intended to annex Indian territory permanently. Rather, China wanted to put India in its place, and having done that, the PLA retreated before winter snows cut off the passes from Tibet.[17]

In numerical terms, it had not been a big conflict. The Indian Army committed only two and a half divisions and lost roughly 7,000 men as casualties or prisoners, while the PLA reported 2,000 casualties. All the fighting took place in the rugged alpine terrain of the eastern and

[17] For a discussion of the different explanations offered for the Chinese withdrawal, see Ramchandra Guha, *India After Gandhi: The History of the World's Largest Democracy* (New York: Ecco, 2007), 339.

western Himalaya, and the Indian Air Force and Navy never engaged in combat. Yet even if the war had been small in life, in the Indian imagination it was an epic—and tragic—confrontation. The names of the dead were chiseled into war monuments across the country. Chetan Anand's film *Haqeeqat* (*Reality*), released two years after the war, told the story of doomed *jawans* battling for their lives in Ladakh. When Bollywood vocalist Lata Mangeshkar sang "*Ae mere vatan ke logon*" (O people of my nation), a mournful song about the martyrs of the war, Nehru was reportedly moved to tears.[18]

The West and India's Air Defenses

Let us leave Nehru to weep undisturbed as we consider the wider implications of the Sino-Indian War. Beyond the depressing movies and sad songs, the war was a crisis that forced lachrymose Nehru and his chastened government to rethink their nation's approach to foreign relations, economics, development, and technology. Autarky and nonalignment as national ideals were shoved to the side as the Indian government focused on military buildup through arms imports.

Even though it had played a strictly noncombatant role in the conflict, the Indian Air Force became a particular focus of the post-1962 militarization. Nehru's November 19 telegrams to Kennedy set the stage for western-bloc involvement in improving India's air defenses. The cities of northern India had no proper air defenses. If the Chinese had decided to escalate the conflict by bombing the cities, the Indian Air Force could have done little to stop them. (This was later cited as the reason why the IAF was not sent to strafe Chinese troops and supply lines, as this could lead to retaliatory bombing.) As if to reinforce the significance of air defenses to Nehru's government, Chinese fighter planes made regular sallies across the McMahon Line after the cease-fire. Although it would have been difficult or impossible for an IAF interceptor to shoot down an

[18] P.B. Sinha, and A.A. Athale, *History of the Conflict with China, 1962* (New Delhi: Ministry of Defence, 1992), http://www.bharat-rakshak.com/LAND-FORCES/Army/History/1962War/PDF/index.html (accessed February 10, 2011), 412; Wortzel, "Concentrating Forces and Audacious Actions," 327; Guha, *India After Gandhi*, 716.

intruder, even with sophisticated radar, the Chinese incursions did their part to inspire public demands for enhanced air defenses.[19]

On invitation from the Indian government, a joint team of military experts from the United States and Commonwealth nations visited India in January and February 1963. The opposition parties and the press criticized the Nehru administration for this visit, arguing that the invitation of a foreign military mission was a violation of nonalignment. Britain, a fading imperial power, presented no danger to India, but some members of the opposition saw the United States as a real threat to Indian sovereignty. An editorial in the *Times of India* warned that the United States and Commonwealth militaries were planning to bring an "air umbrella" to India: foreign fighter planes, based on Indian soil, defending Indian cities. Even though the air umbrella plan was almost exactly what Nehru had requested in his letters to President Kennedy, the Prime Minister flatly denied that the United States–Commonwealth military mission planned to bring such a defense system to India.[20]

Military aid to India also attracted criticism in the West. Like most debates on aid to India and other nonaligned countries, discussions on the Military Assistance Program descended into red-baiting and name-calling, with Nehru appearing as a "weakling in the face of communism" or a "great international phony." A Republican congressman from Texas reported that he had heard that the Kennedy administration was planning to pass some decommissioned radar equipment to India. "If this information is correct, and I believe it is," the congressman declared, "this is a deplorable and almost unbelievable situation, indeed, when our

[19] Memorandum for the Record, November 19, 1962, *Foreign Relations of the United States, 1961–63*, 19:394–96 (hereafter cited as *FRUS* with year and volume); Jasjit Singh, *The Icon: Marshal of the Indian Air Force Arjan Singh, DFC: An Authorised Biography* (New Delhi: KW Publishers, 2009), 121. As then-IAF chief Arjan Singh explained in a 1969 conference paper, aerial incursions were carefully timed so that interceptors could never catch them. A fighter flying at eight miles per minute could cover forty miles of Indian territory before an IAF fighter had the chance to scramble and intercept. If by some fluke an IAF interceptor shot down a Chinese or Pakistani fighter on a regular incursion mission, the international community would doubt that the intruding fighter had been over Indian territory at the time of the shootdown. Arjan Singh, "The Air Force," in *Defence of India* (New Delhi: Press Institute of India, 1969), 29.

[20] *Keesing's Contemporary Archives: Weekly Record of Important World Events with Index Continually Kept Up-to-Date* (London: Keesing's Publications, 1964), 14:19648; "Deceptive," *The Times of India* (*TOI*), July 24, 1963; "Readers' Views," *Indian Express* (*IE*), July 30, 1963.

Government is willing to close down our own military centers and ship our equipment to any country, much less a Socialist country such as India that has continually played footsie with the Communists."[21]

The Military Assistance Program received cautious support from New Frontier Democrats. Chester Bowles, who returned to the New Delhi ambassadorial post in July 1963, had mixed feelings about such aid, as he confided in his memoir *Promises to Keep*. Although he thought that American military aid to non-European countries was usually a bribe for supporting American foreign policy, Bowles felt that India was a special case. Military aid could devolve into a codependent relationship in which the recipient threatened to expel American troops stationed in the country unless aid were increased, but this would not be an issue in India because there was no question of stationing troops in the nonaligned country. Further, if the West did not offer aid, India could turn to the Soviets.[22]

On July 9, outgoing US Ambassador John Kenneth Galbraith delivered to Nehru an official agreement outlining India's and the United States' obligations to each other regarding Indian air defense. The six-point agreement stated that the United States would provide India with permanent radar stations to cover almost the entire northern border. In the year or more that it would take to complete these stations, the United States would lend India two portable radar sets to cover Delhi and Calcutta. The USAF would train Indian crews to operate both the temporary and permanent radar stations, as well as send supersonic fighter squadrons to India intermittently for training exercises with the IAF. In the event of another attack, the United States would ask India if it needed help; if the government of India deemed this to be the case, it would take responsibility for accommodating any American personnel deployed to India.[23]

At the beginning of August, another western military mission arrived in New Delhi to discuss the details of the first joint training exercise. The negotiations established that the IAF-led exercise, held in November of that year, would be commanded by Air Vice-Marshal (AVM) Arjan Singh, Deputy Chief of the Air Staff. The name of the joint exercise

[21] *Cong. Rec.*, 88th Cong., 1st sess., 1963, vol. 109: 15006, A4635.

[22] Chester Bowles, *Promises to Keep: My Years in Public Life, 1941–1969* (New York: Harper & Row, 1971), 476–77.

[23] Telegram from the Embassy in India to the Department of State, July 10, 1963, *FRUS*, 1961–1963, 19:615–17.

would be "Shiksha," after a Sanskrit word for "education."[24] In the exercise, the IAF planned to field Hunter and Gnat fighters as interceptors, and English Electric Canberra bombers as target aircraft. The RAF would send one squadron of Gloster Javelins, while the USAF would dispatch a squadron of North American F-100 Super Sabres. The Royal Australian Air Force also planned to make a token contribution to the exercise, sending thirteen airmen and two of its own Canberra bombers to supplement the IAF fleet.[25]

The war with China had been such a disaster for India that the government was willing to repudiate nonalignment by signing an agreement with a western power and inviting foreign military forces to India for training exercises. But nonalignment had become such an integral part of Indian domestic politics that the air agreement came under criticism as soon as it was announced on July 22. An editorial in *Times of India* condemned the announcement as "sophistry" because of its obfuscation of the air agreement's ramifications concerning nonalignment.[26]

Regardless of the opposition, military planes from the West started to come to India shortly after the conclusion of the air agreement. (They had, in fact, been coming since the previous year, only now they were there to support air defenses rather than the airlift to Ladakh.) In August and September 1963, American transports airlifted two complete radar stations to India. Each temporary station included two radar sets: an AN/MPS-11 search radar and an AN/MPS-16 height-finder (Fig. 5.4). Both radars were portable models that could be deployed and dismantled easily. Together, they would allow participants in Exercise Shiksha to triangulate the positions of the target aircraft. Under the terms of the air agreement, these temporary stations would be replaced by permanent stations. The permanent stations would be previously used, having been

[24] According to Vaman Shivaram Apte, *The Student's Sanskrit–English Dictionary* (Delhi: Motilal Banarsidass, 1970), 554, *śikṣā* means "learning, study, teaching, training, instruction, etc." As used in modern Hindi, *shiksha* means "education." R.S. McGregor (ed.), *Oxford Hindi–English Dictionary* (New Delhi: Oxford University Press, 1993), 950.

[25] "RAF Team Will Arrive Today," *Hindustan Times* (*HT*), August 2, 1963; "Joint Air Exercise: Talks Open for Technical Plan," *HT*, August 3, 1963; "It's Shiksha for IAF," *HT*, August 6, 1963; "U.S. Jets Take Part in Shiksha," *American Reporter*, November 15, 1963.

[26] "Deceptive," *TOI*, July 24, 1963.

Fig. 5.4 American mobile radar sets in use during exercise Shiksha: AN/MPS-11 search (left) radar and AN/MPS-16 height-finder (right) (*Source* "Operation Shiksha," Moving Images Relating to Military Aviation Activities 1909–1984, RG 342, NARA)

passed on to India after the United States upgraded its own defenses. Even with military aid, India still sometimes found itself using secondhand equipment.[27]

The Air Force Gets an Education

On the morning of November 6, 1963, six US Air Force F-100 Super Sabres arrived at Palam airfield in southwest Delhi. Representatives of the Indian press were on hand to watch the sleek, shiny jets fly over the airfield in tight formation, then circle around to land. After the pilots had disembarked from their planes, US Embassy officials and IAF brass gave the visiting pilots a snack of tea and sandwiches as a gesture of hospitality. Ambassador Bowles addressed the pilots in a stirring speech, but Air Marshal Aspy Merwan Engineer, head of the IAF, declined to speak.[28]

[27] "U.S. Radar Equipment Arrives," *HT*, August 17, 1963; "More Radar Equipment Arrives," *HT*, September 16, 1963; General Electric, "Technical Manual Operation: AN/FPS-8 and AN/MPS-11," http://www.mobileradar.org/Documents/MPS-11_TO.pdf (accessed August 14, 2014).

[28] "U.S. Planes Arrive; Stage Set for 'Siksha'," *HT*, November 7, 1963; "U.S. Planes Arrive for India Exercises," *NYT*, November 7, 1963; "U.S. Jets Take Part in Shiksha," *American Reporter*, November 15, 1963; "Operation 'Shiksha'," *Indian Aviation*, November 1963, 319–20, 337; USAF 1365th Photographic Squadron, unedited footage of Exercise Shiksha (35 mm film, reel 1), 1963, US moving images compiled 1947–1984, RG 342, NARA.

The American planes were the first of three waves of fighters belonging to the 354th Tactical Fighter Wing of the Tactical Air Command to come to India for Exercise Shiksha. It had been three days since the pilots had left their home base at Myrtle Beach, South Carolina. With the help of tanker aircraft supplied by the Strategic Air Command, the single-seat planes had crossed the Atlantic and parts of Europe and Asia, stopping at Morón Airbase in Spain and Dhahran Airbase in Saudi Arabia along the way.[29]

Before departing for India, the Tactical Air Command and the American fighter pilots had spent a month preparing for the exercise. To prepare for the mock intercepts they would perform in Exercise Shiksha, the pilots practiced by intercepting Martin B-57 bombers, the American-built version of the English Electric Canberras that the Indian and Australian air forces would use in the exercise. Beyond practical military concerns, the pilots also learned about social and political conditions in India by watching US Army "Area Study" films and attending briefings by intelligence officers. As the final report of the 356th Tactical Fighter Squadron declared, "Every effort was made to impress upon all participants in the exercise the importance of making a good impression for Tactical Air Command, the United States Air Force and the United States of America."[30]

After arriving in India, the pilots received an additional briefing from Brigadier General Gordon M. Graham, who warned them of sensitive topics to avoid in discussions with Indian airmen. Graham was the commanding officer of the American forces in the exercise, working opposite Air Vice-Marshal E.W. Pinto, who commanded the Indian forces in the western sector. (Indian forces in the eastern sector, where the British squadron deployed, had their own commanding officer, Air Vice-Marshal Shiv Dev Singh.) Judging from Graham's own account in his memoir *Down for Double*, the Brigadier General grouched his way through the exercise, grumbling about the IAF's fleet of "real trash for airplanes" and seeing communists around every corner in Delhi. Although Graham purportedly got along famously with Pinto, he spent hours arguing with customs officials upon landing at Palam. He even clashed with Chester

[29] History of the 354th Tactical Fighter Wing (TAC), July–December 1963, Call # K-Wg-354-HI, IRIS # N0728, in the USAF Collection, AFHRA; History of the 499th Air Refueling Wing (SAC), November–December 1963, Call # K-Wg-499-HI, IRIS # 00463290, ibid.

[30] History of the 354th Tactical Fighter Wing (TAC), July–December 1963.

Bowles. When Graham visited the embassy in uniform and found Bowles wearing an untucked sport shirt, he seems to have wondered if the ambassador had gone native.[31]

In addition to the F-100 pilots themselves, US Air Force and Navy ground crew, radar operators, publicists, and command officers deployed to India for Exercise Shiksha, arriving with the necessary supplies and support equipment in transport planes. About 300 American military personnel were housed in Camp Shiksha, located in a secure area of Palam airfield. Like an American embassy housing compound, Camp Shiksha was a little colony that reproduced, as closely as possible, the conditions and culture of the home country. Fresh food was flown in from Incirlik Air Base in Turkey. Six US Air Force cooks prepared meals for the American personnel stationed at Camp Shiksha. The all-American menus included such dishes as ham, roast beef, and southern fried chicken.[32]

The American airmen who participated in the exercise did not spend their entire time in the colony of Camp Shiksha. They took time out for some predictable tourist fare, making a day trip to Agra to see the Taj Mahal. A USAF camera crew that accompanied one group to Agra shot footage of the airmen and their guide admiring Shah Jahan's magnificent mausoleum, riding in a horse-tonga, and making obligatory stops at handicraft emporia. At some point in their deployment to India, the airmen were treated to a garden party and cultural program, along with some of New Delhi's well-dressed elites. The party guests ate, drank, chatted, and watched groups of girls and boys perform traditional dances.[33]

The American airmen's main duty in India was participating in the joint air exercise, and it was on the exercise that they spent the majority of their time in India. Exercise Shiksha was split into two sectors, eastern (RAF) and western (USAF), each equipped with its own portable American radar station. The RAF Javelin fighters, which arrived in India more than a week ahead of the USAF fighters, were stationed at IAF Kalaikunda near

[31] History of the 354th Tactical Fighter Wing (TAC), July–December 1963; "Joint Exercises Will Train 4,000 IAF Personnel," *HT*, October 25, 1963; Gordon M. Graham, *Down for Double: Anecdotes of a Fighter Pilot* (White Stone, VA: Brandylane Publishers, 1996), 167–69.

[32] "Six Cooks Feed 300 in Shiksha," *Air Force Times*, November 27, 1963.

[33] "Air Training Exercises Prove Valuable," *American Reporter*, November 29, 1963; USAF 1365th Photographic Squadron, unedited footage of Exercise Shiksha (35 mm film, reels 2 and 10), 1963, US moving images compiled 1947–1984, RG 342, NARA.

Calcutta. The IAF and RAAF target bombers were stationed at the airbase in Agra.[34]

The nerve center of the exercise was the tracking station, housed inside a Quonset hut next to the radar antennas. Inside the darkened hut, radar operators hunched over screens displaying signals from the AN/MPS-11 search radar as its antenna twirled around and around outside. The operators were watching for blips indicating the approach of target bombers. When a bomber appeared on the radar, the operators used the height-finding AN/MPS-16 antenna to triangulate its location. Then the commander telephoned an alert to the fighter pilots, who were waiting in a ready room at the airbase. When the pilots received the signal, they dashed off to their planes on the flightline to initiate a scramble—a quick takeoff to intercept the target bombers. The fighters then navigated to the point in the sky where the radar said the bombers would be—and if all went well, the bombers would be there.[35]

All did not go well on the first day of Exercise Shiksha, which began in the eastern sector on November 9. Indian and British pilots spent the first day trying to intercept the target bombers, which flew within a 1,000-square-mile area between 20,000 and 50,000 feet elevation. Even by following the directions issued by the tracking station, the defending aircraft were able to intercept disappointingly few of the attacking planes.[36]

Exercise Shiksha began in the western sector on November 14, which was also the last day of the exercise for the eastern sector. About 140 Indian, British, and American fighter aircraft took to the air that day, making it the busiest day of the exercise. The fighters flew practice interceptions against fifty Canberra bombers. The exercise zone stretched across north India, from West Bengal in the east to Punjab in the west.

[34] "Joint Exercises Will Train 4,000 IAF Personnel," *HT*, October 25, 1963; "RAF Group Arrives for Exercises," *HT*, October 18, 1963; "British Planes to Join Indian Exercises," *The Times* (London), October 16, 1963.

[35] USAF 1365th Photographic Squadron, unedited footage of Exercise Shiksha (35 mm film, reels 6 and 10), 1963, US moving images compiled 1947–1984, RG 342, NARA. According to the *Oxford English Dictionary*, the term "scramble" was first used in this context by the RAF in the Battle of Britain.

[36] "'Attacking' Planes Give Good Account in Shiksha," *HT*, November 10, 1963; "India's Air Defense Fails in First Drill," *NYT*, November 10, 1963.

From time to time during the day, residents of Delhi caught sight of fighters and bombers above the city.[37]

In both sectors, the exercise took place in two phases. After two or three days of practice intercepts, all exercise participants would stand down for a day for evaluation and critique of their performance. In the western sector, IAF and USAF pilots flew missions out of Palam on November 14–15 and 17–19. On November 16, commanding officers reviewed exercise data and offered suggestions for improvement. One of the types of data reviewed was gun-camera photographs. Pilots recorded their interceptions by exposing a frame of film with small cameras mounted in their aircraft. As Wing Commander Donald Michael, the commanding officer of IAF No. 3 Wing, recalled, the IAF pilots had taken several good pictures of the target bombers, but the films from the American planes were blank. The Indian planes were armed with cannons, but the American planes would, in real combat, carry Sidewinder missiles. As soon as their planes' radar locked on to the target aircraft, the American pilots exposed their gun-camera film, because they could trust their missiles to find and destroy targets beyond visual range.[38]

All the hardware used by the IAF in the exercise was of western-bloc origin, and the Indians took care to keep Soviet technology out of sight. This was common practice in India during the economic Cold War: segregate Americans and Soviets so they would never see each other or the other's technology, even though everyone knew that India was actively courting both sides. On one occasion in 1960, a delegation of American atomic scientists was whisked away to the Taj Mahal immediately after landing in Delhi, to avoid their running into a rival Soviet group visiting at the same time.[39]

In Exercise Shiksha, the IAF's supersonic MiG-21s remained at an airbase separate from the exercise. When a member of the Indian press asked a Defence Ministry spokesman why the MiGs had been excluded from the exercise, the spokesman replied obtusely, "MiGs are not being

[37] "140 Jet Fighters Take Part in 'Air Battle'," *HT*, November 15, 1963.

[38] History of the 354th Tactical Fighter Wing (TAC), July–December 1963; Donald Michael, "Exercise Shiksha," *Bharat Rakshak*, http://www.bharat-rakshak.com/IAF/History/1960s/Shiksha01.html (accessed February 13, 2011).

[39] David C. Engerman tells the story of the surprise visit to the Taj Mahal in *The Price of Aid: The Economic Cold War in India* (Cambridge, MA: Harvard University Press, 2018), 1.

used because we do not want to mix up the two." As Chapter 7 will discuss in more detail, US policy explicitly forbade the use of eastern-bloc hardware in American-financed development projects. The July 1963 air agreement was not subject to these same restrictions, but MiG-21s were kept away from Exercise Shiksha because the Indian defense establishment understood that the Americans would not appreciate a reminder of India's connections with the Soviet Union.[40]

By excluding the MiG-21, the Indians were also respecting Soviet desires for secrecy. The Americans had a natural curiosity about the MiG-21, the most sophisticated Soviet fighter in operation at the time. In his memoir, the ever-cantankerous Brigadier General Graham claims that he was able to see inside the cockpit of a MiG-21 with the help of AVM Pinto. Posing as an exchange officer, Graham went to the airbase in an IAF uniform, where a pilot officer checked him out in the MiG-21. When Graham was preflighting the aircraft, within minutes of taking off, the pilot officer ran back out to the flightline and told Graham that he had to leave immediately. Despite Pinto's precautions, word of Graham's attempt to fly the MiG-21 had leaked to the IAF high command, who put a stop to the flight before it got off the ground.[41]

As the Americans snuck peeks at the MiG-21s, the western-bloc planes were visible to the Indian press and government. The Indians, for their part, were impressed by the high-performance western fighters. In the press, the RAF Javelins elicited some admiration for their all-weather capabilities, but the supersonic USAF F-100 Super Sabres received the most attention. An article in the conservative and generally pro-western newspaper *The Hindu* reported that the F-100 could attain supersonic speeds in level flight while carrying a fearsome array of weapons. "The F-100's are really fantastic fighters," the article concluded.[42]

Like the press, the IAF leadership was also enchanted by the western-bloc fighters. The air force brass' preference for western fighters was heightened by jealousy of Pakistan, which had operated the F-104 Starfighter since 1961. As events in the 1971 Indo-Pakistani War would

[40] "Shiksha Improves IAF Interception Techniques," *HT*, November 17, 1963.

[41] Graham, *Down for Double*, 173–74.

[42] "Better Show by Indian & U.S. Pilots," *The Hindu*, November 14, 1963.

show, India's MiG-21s were more capable fighters than Pakistan's F-104s.[43] At the time of Exercise Shiksha, though, bigger and more complex American fighters held an appeal to the Indians that the Soviet fighters lacked.

November 19 was the last day of Exercise Shiksha, with a final round of practice intercepts performed in the western sector. General Walter J. Sweeney, head of the US Tactical Air Command, visited India on the last day of the exercise. (Predictably, Brigadier General Graham clashed with General Sweeney on his short visit.) The general told the press that "cooperation between the IAF, USAF, RAF and Australian Air Force appears to have been superb." The Indian Ministry of Defence also released a communique expressing its satisfaction with the training value of Exercise Shiksha, which gave the IAF "valuable experience" in operating radar equipment under realistic conditions.[44]

Friends and Rivals in Exercise Shiksha

Behind the bonhomie, there had been some tensions in East–West—especially Indo-American—relations during Exercise Shiksha. Throughout the exercise, Indians and Americans misunderstood each other. These misunderstandings could be simple communication errors, such as when the driver of an ambulance posted at the Palam flightline drove away at 1:30, when the day's flight operations were only half complete. ("Although we were supplied with an ambulance by the Indian Air Force for the coverage of crash calls," the final report of the 356th Tactical Fighter Squadron explained with evident frustration, "the situation was far from desirable.")[45] Misunderstandings could also be deeper and more intractable, based on one culture's perceptions of the other.

Exercise Shiksha took place in the early sixties, almost two decades after the American encounter with India in World War II and before the opening of the hippie trail to the exoticized East later in the decade. Americans who participated in the exercise brought along stereotypes of India as a disorderly and unpredictable place—perceptions that

[43] Pushpindar Chopra, "India and the MiG-21," *Air Enthusiast*, July 1973, 10.

[44] History of the 354th Tactical Fighter Wing (TAC), July–December 1963; "Air Exercise Ends: RAF Unit Leaves," *HT*, November 20, 1963; "I.A.F. Gains Good Experience," *TOI*, November 21, 1963; "Air Tests in India Said to Open Way for Fast U.S. Aid," *NYT*, November 22, 1963; Graham, *Down for Double*, 172–73.

[45] History of the 354th Tactical Fighter Wing (TAC), July–December 1963.

had affected Americans' relationships with India and Indians ever since Katherine Mayo's *Mother India*. Brigadier General Graham's trash-talking the IAF fleet may be as much an outcome of his disagreeable personality as it was of a deeply internalized Orientalism. Nevertheless, Graham was taking part in a long western tradition of negatively judging non-western people because of their technology.[46]

For their part, Indians also had their own stereotypes of Americans and their country. Some perceptions were positive, like Nehru's evaluation of the United States as a country possessing a "vital energy" that most of the rest of the world lacked. But many perceptions were negative. The Cold War and civil rights movements loomed large in Indian perceptions of America as a discriminatory society and an international bully. And then there was Hollywood's own inadvertent propagandizing with its endless supply of cowboy and gangster movies. Without much else to base their perceptions on, the Indians who watched exported versions of these films thought that real American men were as arrogant and violent as the characters in their movies.[47]

Bad stereotypes of Indians lived on in American memory of the American Military Assistance Program. Nowhere is this more apparent than in an account written by aviation reporter Joseph E. Dabney in 1979. An account of the USAF airlift into the Himalayan region appears in Dabney's book-length biography of the C-130 Hercules transport aircraft. Indians appear in the account only as foils; the real characters are the plane itself and the American men who fly it. This is the style of the entire book, which is fittingly titled *Herk [Hercules]: Hero of the Skies*. From the first pilot of the prototype YC-130, "gutsy flier of the old school" Stanley Beltz, to the USAF pilots who shuttled supplies up to Ladakh and NEFA, heroic American men flew the heroic Hercules on missions of masculinity.[48]

[46] Graham, *Down for Double*, 169; Michael Adas, *Machines as the Measure of Men: Science, Technology, and Ideologies of Western Dominance* (Ithaca, NY: Cornell University Press, 1989), 4, 402; Andrew J. Rotter, *Comrades at Odds: The United States and India, 1947–1964* (Ithaca, NY: Cornell University Press, 2000), 18, 85.

[47] Rotter, *Comrades at Odds*, 24, 169. For international coverage of the American civil rights movement, see Mary L. Dudziak, *Cold War Civil Rights: Race and the Image of American Democracy* (Princeton, NJ: Princeton University Press, 2000).

[48] Joseph E. Dabney, *Herk: Hero of the Skies* (3rd ed., Fairview, NC: Bright Mountain Books, 2003), 79, 152–60.

Compared to the Americans, Indians in Dabney's account are little more than children. Thinking of non-western peoples as children was a staple of colonial Orientalism. This interpretation persisted in American international diplomatic and development relations in the Cold War (and judging from this author's conversations with some veterans of the twenty-first-century wars in Afghanistan and Iraq, it lives on among some members of the US armed forces to this day). Dabney's informants, and likely also Dabney himself, had imbibed these ideas, thus affecting the narratives they told. A recurring character in Dabney's tale is an anonymous Indian air vice-marshal, who at every turn in the narrative is overcome in childlike amazement by the capabilities of the American planes. According to one USAF transport pilot quoted by Dabney, *jawans* were shuttled to airfields in Assam, 120 seated at a time on the floor of a C-130's cargo hold. The *jawans* sat obediently still during the flight, "but when we landed and they filed out, you could see where many of them had wet their pants."[49]

Dabney's informants had a dim view of Indian planes as well as people. The USAF C-130s operated alongside An-12s operated by the IAF. The Soviet-built An-12 was a cargo plane powered by four turboprop engines, similar in configuration to the C-130. Both planes had similar ranges and payload capacities. The An-12 had a faster maximum speed than the C-130, but it lacked pressurization in the cargo compartment.[50] One American pilot claimed that the An-12 cockpit was a rat's nest of exposed wiring. "In the cockpit, you almost had to be a cat-walker to get around the wiring," he is quoted as saying. "There was hardly any flooring and cables were all around you."[51] Even in the high-tech world of air transportation, Americans saw India as disorderly and unpredictable.

The same events are perceived from a rather different perspective in Indian memories. In 2002, retired IAF An-12 pilot Group Captain A.G. Bewoor wrote a response to Dabney's account for the Indian website *Bharat Rakshak*. Even though Dabney's account is a blatant example of ethnocentrism, an introduction by the webmaster simply states that Dabney had included "a certain amount of hype" in his account, which

[49] Dabney, *Herk*, 152–55; Rotter, *Comrades at Odds*, 85–86.

[50] Range: 2,113 mi (An-12), 2,090 mi (C-130); payload capacity: 44,000 lb (An-12), 45,000 lb (C-130); maximum speed: 444 mph (An-12), 383 mph (C-130). René J. Francillon, *Lockheed Aircraft Since 1913* (Annapolis: Naval Institute Press, 1987), 376; Singh, *Aircraft of the Indian Air Force*, 134.

[51] Dabney, *Herk*, 159.

Bewoor was endeavoring to correct. The American pilots were not the pioneers that Dabney made them out to be. IAF An-12s had been flying the same routes through the Himalaya before the USAF planes arrived, and IAF Dakotas and Packets had been flying to Leh since Partition.[52]

Responding to Dabney's dismissal of the An-12, Bewoor concedes that the Russian-made plane was inferior to the C-130; yet it was a capable design in its own right. "While is it true that the cargo compartment of the AN12 is not pressurized," he writes, "it is absolute poppycock to add that 'hauling troops over the Himalayas was completely out of the question.' What did the AN12s do for 32 years if not hauling troops across the Himalayas? Do the Herk pilots think we just took cabbage and chicken?" As for the An-12 cockpit's being a maze of wires, the American pilot must have visited a plane in a maintenance hangar, not an active aircraft on the flight line.

In the Military Assistance Program and Exercise Shiksha, points of friction between Indians and Americans ranged from small-scale errors of translation like the absconding ambulance driver to more fundamental misunderstandings stemming from ethnocentrism and Orientalism. Big and small, these misunderstandings did not hinder the Military Assistance Program. Yet they are indicative of the wider Cold-War tensions between the two countries. Andrew J. Rotter notes that "decision makers are creatures of culture, not just policy wonks who shed their images of others like raincoats at the office door."[53] The same could be said of the Indian and American airmen who participated in Exercise Shiksha. Although Indians appreciated the military aid offered by the United States, they disliked cavalier American attitudes about Indian culture and technology. There is ample evidence of respect and even camaraderie between Indian and American individuals during the exercise, but American policy approached India as a younger and weaker nation. Helping India to defend itself against communists was part of the "white man's burden" that Americans inherited from European colonialism and now carried as the leading western power.

[52] A.G. Bewoor, "The Airlift into Leh," *Bharat Rakshak*, http://www.bharat-rakshak.com/IAF/history/1962war/1263-leh-airlift.html (updated June 12, 2017).

[53] Rotter, *Comrades at Odds*, xx.

From Palam Field to Kanjarkot Fort

Two days after the end of Exercise Shiksha, the USAF F-100 Super Sabres of the 354th Tactical Fighter Wing departed Palam Field and headed for their home base at Myrtle Beach. The end of the exercise was overshadowed in the Indian press by dual tragedies. On the same day as the departure of the American fighters, a helicopter crash in the mountains of Jammu and Kashmir claimed the lives of five high-ranking officers of the Indian military, including AVM Pinto. Pictures of the five deceased were printed on the bottom of the front page of the next day's *Hindustan Times*. The top headline for the day, though, announced the assassination of President John F. Kennedy in Dallas. The pioneer of the New Frontier was dead at 46. When the American fighter pilots returned to their home base at Myrtle Beach on November 25, they joined the rest of the US Air Force in a month-long period of formal mourning.[54]

Exercise Shiksha had ended and would soon be forgotten. The footage shot by USAF camera crews during the exercise seems never to have been used for anything. (It eventually ended up in the vaults of the National Archives in College Park, Maryland.) There was never a second exercise, even though the July 1963 agreement had stated that these exercises would take place periodically for the strengthening of India's air defenses.[55] In planning for this one and only exercise, the United States and Britain had hoped that their military aid programs to India would at once discourage further Chinese ambitions in South Asia and strengthen ties between India and the West. India merely wanted modern radar and the training to use it effectively. Having gotten what it wanted with Exercise Shiksha, India felt no need to call for another joint air exercise.

Policymakers in the United States continued to view India as a valuable de facto ally living under constant threat from communist China. In December 1963, General Maxwell Taylor visited India for talks with government officials. Chairman of the Joint Chiefs of Staff for President Kennedy and now President Johnson, General Taylor met with Prime

[54] "Pinto, Daulet Singh, 3 Other Officers Die in Copter Crash" and "President Kennedy Is Assassinated," *HT*, November 23, 1963; History of the 3974th Combat Support Group (SAC), November 1963, Call # K-Gp-Sup-3974-HI, IRIS # M0157, in the USAF Collection, AFHRA; History of the 354th Tactical Fighter Wing (TAC), July–December 1963.

[55] *Keesing's Contemporary Archives*, 14:19648; "Use of Radar: I.A.F. Gain Good Experience," *The Hindu*, November 17, 1963.

Minister Nehru, Defense Minister Chavan, Finance Minister T.T. Krishnamachari, and the heads of the service branches. He also visited Indian military installations in Ladakh and NEFA. In his final report on his visit to India, he recommended that the United States indicate willingness to offer $50 million of interim military aid on the condition that India undertake longer-term military planning.[56]

Ten months after General Taylor's tour of India, China became an exponentially bigger threat when, on October 16, 1964, the Chinese nuclear establishment detonated its first atomic bomb at the Lop Nur Nuclear Weapons Test Base in Sinkiang (Xingjiang). The Chinese had begun developing their bomb with substantial Soviet assistance, although this aid was cut off in April 1960 with the collapse of Sino-Soviet relations. The USSR had also aided China's missile program by shipping two R-2 (SS-2) missiles to China in January 1958. By 1960, Chinese engineers had already constructed a copy of the R-2 and were working on modifying the design. As a result, in 1964, not only did China have nuclear weapons, the country was also actively developing the ability to deliver these weapons to neighboring countries such as India.[57]

The United States saw the Chinese bomb test as another opportunity to expand Indo-American cooperation. "We believe our focus with [the] Indians should be on [the] problem of building [the] prestige of Indian science and technology," Under Secretary of State George Ball reported in a telegram to the embassy in Delhi two months after the first Chinese bomb test. Providing direct assistance to an Indian nuclear weapons program was out of the question, and in any event India did not yet have such a program. But the US government saw plenty of opportunity to expand nonmilitary Indian science, and by early 1965, talks with Indian scientists such as Homi J. Bhabha had already begun.[58]

[56] Maxwell Taylor to Robert McNamara, December 23, 1963, *FRUS*, 1961–1963, 19:722–26; "Balanced Defence Program," *American Reporter*, December 27, 1963.

[57] John Wilson Lewis and Xue Litai, "Strategic Weapons and Chinese Power: The Formative Years," *China Quarterly* 112 (December 1987): 541–45, 549; Asif A. Siddiqi, *Challenge to Apollo: The Soviet Union and the Space Race, 1945–1974* (Washington, DC: NASA History Division, 2000), 57–58, 290. The R-2 missile was a modified and upgraded version of the R-1, which in turn was a Russian copy of the V-2 missile from Nazi Germany.

[58] Telegram, George W. Ball to Embassy in India, December 12, 1964, *FRUS*, 1964–1968, 25:169–71; "Indian Nuclear Energy Program," memo, February 22, 1965, ibid., 187–90.

In April 1965, conflict broke out again in the Indian subcontinent. This time, India's adversary was not China but the American ally in South Asia, Pakistan. Indian and Pakistani ground forces began skirmishing around Kanjarkot Fort in the Kutch wasteland between Gujarat and Sind. At the beginning of September, the conflict erupted into a full-blown, albeit brief, war. India used a Pakistani infiltration of Kashmir as a pretext for counterattacks in Pakistan-controlled Kashmir, Punjab, and Sind. The two countries fought to a stalemate over the next three weeks. A UN-brokered cease-fire took effect on September 22, 1965. The following January, India and Pakistan negotiated the formal end of the conflict at Tashkent in the USSR, with Soviet Premier Alexei Kosygin mediating.[59]

In the 1965 war Pakistan, and to a lesser extent India, used military equipment and training that had been provided by the United States. Rather than equipping India and Pakistan to fight against communists, American military aid had armed the two countries to go to war against each other. On September 8, 1965, as the conflict was rapidly escalating, the Johnson administration announced that it would cut off all military aid to both India and Pakistan.[60]

Conclusion

In several respects, the middle years of the 1960s represent a turning point in the history of modern India. The 1962 conflict with China discredited Nehru's belief that a nonaligned India could lead Asia by moral force rather than military might. The conflict also complicated India's efforts to attain economic and technological self-reliance. Nehru's trickle-down argument, that an expansion of military production capacity must naturally follow the growth of civilian industry, now seemed inadequate. Industrial autarky could only come slowly, but India needed more military equipment immediately. To get these arms, India would have to import them. After the wars with China and Pakistan, India ramped up military expenditure. Between 1962–1963 and 1963–1964, India's total reported military expenditures jumped from 4.74 to 8.16 billion rupees ($1 to

[59] Guha, *India After Gandhi*, 398–405; McMahon, *Cold War on the Periphery*, 327–31. For a representatively propagandistic account of the 1965 war, see Hari Ram Gupta, *India–Pakistan War 1965* (2 vols., Delhi: Hariyana Prakashan, 1967).

[60] McMahon, *Cold War on the Periphery*, 328–29.

$1.7 billion). By 1969–1970, India's military budget had risen to 11.83 billion rupees ($890 million).[61]

As a nonaligned nation, India sought to receive military aid from the western and eastern blocs at the same time. Yet as the American Military Assistance Program to India illustrates, importing arms—whether by direct purchase or through subsidized military aid—was contingent on many factors. When it came to American aid to India, domestic opposition in both countries had the effect of limiting aid. Political opposition in the US government opposed sending military aid to nonaligned India. At the same time, opposition within the Indian government objected to the political consequences that might entail from accepting aid from the United States.

The Indian defense establishment regularly met with representatives of both the United States and the USSR. When Defence Minister Y.B. Chavan met Soviet Ambassador Ivan Benediktov in June 1964, he informed the ambassador about his government's aid agreements with the United States. Three months later, in turn, Chavan explained the MiG-21 deal to Ambassador Bowles. In the earlier meeting with Chavan, Benediktov accused the United States of waffling in its aid decisions, while the Soviet Union had always been decisive. What Benediktov did not acknowledge was that the USSR could afford to be decisive because its single-party political system suppressed opposition; the power-sharing American system permitted Congress to reduce or block a President's aid package.[62]

Throughout the negotiations for the sale of fighters and other military assistance to India, American diplomats had repeatedly exhorted the Indians not to rely on the Soviets for aid, because they could easily reach rapprochement with China and then cut off aid to India as a result. Yet

[61] John W. Garver, "India, China, the United States, Tibet, and the Origins of the 1962 War," *India Review* 3, no. 2 (April 2004): 171; Shamlal, ed., *The Times of India Directory & Yearbook 1967* (Bombay: Times of India Press, 1967), 7; Sham Lal, ed., *The Times of India Directory & Yearbook 1969* (Bombay: Times of India Press, 1969), 12; Sham Lal, ed., *The Times of India Directory & Yearbook 1971* (Bombay: Times of India Press, 1969), 2. The US dollar conversion for the 1969–1970 figure is lower than the previous figures because of the devaluation of the rupee in 1966, from 4.76 to 7.50 rupees to the dollar.

[62] P.V.R. Rao, memo about discussion with Soviet Ambassador, June 11, 1964, S. No. 18, Ambassador to USA subject files, papers of B.K. Nehru, NMML; P.V.R. Rao, memo about meeting with Chester Bowles, September 18, 1964, ibid.

as the sixties progressed, Sino-Soviet relations grew ever more acrimonious, at last blowing up in a border war in 1969. In the end, it was not the Soviets but the Americans who proved to be the fickle providers of military aid.[63]

The United States was a definite loser in the Cold-War geopolitical game in South Asia of the 1960s. By allying with Pakistan, the United States had alienated India. The Military Assistance Program to India, in turn, alienated Pakistan. And the United States' yanking away of military aid in September 1965 created even deeper resentment on both sides of the Radcliffe Line. After the Rann of Kutch incident, the United States was already so bankrupt of political clout that it could do nothing but watch Britain negotiate an Indo-Pakistani cease-fire. After the full war, the Soviet Union moved in to occupy the vacuum left by the American geopolitical failure in South Asia. Kosygin appeared as a peacemaker, while the United States had served as a warmonger by arming both belligerents.[64]

American political and strategic influence in India was greatly curtailed after 1965, and it would not be until after liberalization and the end of the Cold War that the Indian and American militaries would have much to do with each other again. But outside of the military and high politics, American influence in India continued to be significant. A bill to block all aid to both India and Pakistan failed to pass the US Congress in early September 1965, and civilian development aid continued uninterrupted after the war.[65] Civilian development—whether US-assisted or not—was as much a part of Indian programs for self-reliance and autarky as was militarization, and it is accordingly the focus of the rest of this book.

Before we leave the Indian Air Force behind and move on to the civilian infrastructure and industry of Nehruvian India, there is one more lesson to consider from the US Military Assistance Program. Exercise Shiksha was a site of technology transfer, both in terms of artifacts and expertise. The transfer was mainly unidirectional, although it is

[63] P.V.R. Rao, memo about meeting with Chester Bowles, September 18, 1964, S. No. 18, Ambassador to USA subject files, papers of B.K. Nehru, NMML; Joseph L. Nogee and Robert H. Donaldson, *Soviet Foreign Policy Since World War II* (3rd ed., New York: Pergamon Press, 1988), 238.

[64] McMahon, *Cold War on the Periphery*, 325.

[65] McMahon, *Cold War on the Periphery*, 329.

possible that the US Air Force's experience in India informed the service's operations in southeast Asia later in the 1960s.[66]

From the United States, India received modern air defense radar and training for how to use it. It is from these radar sets that we learn our final lesson for this chapter. The radar sets, more than any other part of the Military Assistance Program, demonstrate how nonalignment complicated India's defense modernization. The portable radar sets used in Exercise Shiksha were for temporary use only, until permanent secondhand radar stations could be installed and brought online. When the United States cut off military aid in September 1965, the permanent stations remained incomplete. To bring the stations online, the IAF had to improvise, making do with components that were available at the time in India. Even when importing military equipment from one of the world's great military and industrial powers, improvisation was still sometimes necessary.[67]

Throughout the early-independence period, India's access to technology was often affected by international and regional politics. In the case of the radar stations, international politics at first facilitated India's access to the technology, but then later those same politics prevented India from importing the parts to finish constructing the stations. In this way, the Cold War, along with economics and domestic politics, influenced and often frustrated Indian programs of technological modernization and indigenization.

References

1. Adas, Michael. *Machines as the Measure of Men: Science, Technology, and Ideologies of Western Dominance*. Ithaca, NY: Cornell University Press, 1989.
2. Bewoor, A.G. "The Airlift into Leh." *Bharat Rakshak*. http://www.bharatrakshak.com/IAF/history/1962war/1263-leh-airlift.html (updated June 12, 2017).

[66] An article in the *New York Times* at the close of the exercise mentioned another possible example of knowledge gained by the United States in India. The article states that Americans learned "certain operational techniques" from the IAF, which the USAF might adopt. One of these was broadcasting mission instructions to pilots by loudspeaker while they are running to their planes during a scramble, rather than delivering the instructions by radio after the pilots have boarded their aircraft. "Air Tests in India Said to Open Way for Fast U.S. Aid," *NYT*, November 22, 1963.

[67] Singh, *The Icon*, 124.

3. Bowles, Chester. *Promises to Keep: My Years in Public Life, 1941–1969*. New York: Harper & Row, 1971.
4. Chopra, Pushpindar. "India and the MiG-21." *Air Enthusiast* (July 1973): 7–14.
5. Dabney, Joseph A. *Herk: Hero of the Skies*. Fairview, NC: Bright Mountain Books, 2003.
6. Devereux, David R. "The Sino-Indian War of 1962 in Anglo-American Relations." *Journal of Contemporary History* 44, no. 1 (2009): 71–87.
7. Engerman, David C. *The Price of Aid: The Economic Cold War in India*. Cambridge, MA: Harvard University Press, 2018.
8. Francillon, René J. *Lockheed Aircraft Since 1913*. Annapolis: Naval Institute Press, 1987.
9. Garver, John W. "India, China, the United States, Tibet, and the Origins of the 1962 War." *India Review* 3, no. 2 (April 2004): 171–82.
10. Gopal, Sarvepalli. *Jawaharlal Nehru: A Biography*. 3 vols. Cambridge, MA: Harvard University Press, 1984.
11. Graham, Gordon M. *Down for Double: Anecdotes of a Fighter Pilot*. White Stone, VA: Brandylane Publishers, 1996.
12. Guha, Ramachandra. *India After Gandhi: The History of the World's Largest Democracy*. New York: Ecco, 2007.
13. Gupta, Hari Ram. *India–Pakistan War 1965*. 2 vols. Delhi: Hariyana Prakashan, 1967.
14. *Keesing's Contemporary Archives: Weekly Record of Important World Events with Index Continually Kept Up-to-Date*. London: Keesing's Publications, 1964.
15. Lal, Sham, ed. *The Times of India Directory & Yearbook 1971*. Bombay: Times of India Press, 1969.
16. Lal, Sham, ed. *The Times of India Directory and Yearbook Including Who's Who 1969*. Bombay: Times of India Press, 1969.
17. Lewis, John Wilson, and Xue Litai. "Strategic Weapons and Chinese Power: The Formative Years." *China Quarterly* 112 (December 1987): 541–54.
18. Liu, Xuecheng. *The Sino-Indian Border Dispute and Sino-Indian Relations*. Lanham, MD: University Press of America, 1994.
19. Maxwell, Nevill. *India's China War*. London: Cape, 1970.
20. McMahon, Robert J. *The Cold War on the Periphery: The United States, India, and Pakistan*. New York: Columbia University Press, 1994.
21. Michael, Donald. "Exercise Shiksha." *Bharat Rakshak*. http://www.bharat-rakshak.com/IAF/History/1960s/Shiksha01.html (accessed February 13, 2011).
22. Moraes, Frank. *Jawaharlal Nehru: A Biography*. New York: Macmillan, 1956.

23. Nogee, Joseph L., and Robert H. Donaldson. *Soviet Foreign Policy Since World War II*. 3rd ed. New York: Pergamon Press, 1988.
24. Raghavan, Srinath. "A Bad Knock: The War with China, 1962." In *A Military History of India and South Asia, from the East India Company to the Nuclear Era*, edited by Daniel Marston and Chandar S. Sundaram, 157–74. Westport, CT: Praeger Security International, 2007.
25. Raghavan, Srinath. *War and Peace in Modern India*. New Delhi: Permanent Black, 2010.
26. Rotter, Andrew J. *Comrades at Odds: The United States and India, 1947–1964*. Ithaca, NY: Cornell University Press, 2000.
27. Shamlal, ed. *The Times of India Directory & Yearbook 1967*. Bombay: Times of India Press, 1967.
28. Siddiqi, Asif A. *Challenge to Apollo: The Soviet Union and the Space Race, 1945–1974*. Washington, DC: NASA History Division, 2000.
29. Singh, Arjan. "The Air Force." In *Defence of India*, 23–32. New Delhi: Press Institute of India, 1969.
30. Singh, Jasjit. *The Icon: Marshal of the Indian Air Force Arjan Singh, DFC: An Authorised Biography*. New Delhi: KW Publishers, 2009.
31. Singh, Pushpindar. *Aircraft of the Indian Air Force, 1933–73*. New Delhi: English Book Store, 1974.
32. Sinha, P.B., and A.A. Athale. *History of the Conflict with China, 1962*. New Delhi: History Division, Ministry of Defence (Government of India), 1992.
33. Varma, Shanti Prasad. *Struggle for the Himalayas: A Study in Sino-Indian Relations*. Jullundur, India: University Publishers, 1965.
34. Wortzel, Larry M. "Concentrating Forces and Audacious Actions: PLA Lessons from the Sino-Indian War." In *The Lessons of History: The Chinese People's Liberation Army at 75*, edited by Laurie Burkitt, Andrew Scobell, and Larry M. Burkitt, 327–52. Carlisle, PA: Strategic Studies Institute, 2003.

CHAPTER 6

Spanning Assam's Great River: The Brahmaputra Bridge and the Indian Railways, 1958–1962

If you ride a train from Kolkata in West Bengal to Guwahati in Assam, you will be traveling on a route that did not exist before independence and Partition. Your train leaves the bustling red-brick Howrah Station in the early evening. As the train speeds northward outside of the city, village ponds and tanks alongside the tracks reflect the last light of the day. You turn in on your bunk in sleeper-class and are sound asleep as the train passes through the Siliguri Corridor or Chicken's Neck, a narrow strip of land that connects India's northeastern states with the rest of the country. These rail lines through the corridor were built postindependence, after the old lines to Assam were cut off by the new international border with East Pakistan.

When you wake up, your train is in North Bengal and heading eastward toward Assam. As the morning wears on, the train crosses into Assam and passes mile after mile of rice fields. The wide-open spaces of Lower Assam seem to go on forever. To the south, the dramatic forms of the Garo Hills emerge out of the haze, at length merging into the Khasi Hills. The train makes a long turn toward the south, its wheels squealing against the tracks. Then all of a sudden, you are clattering over a bridge, the murky waters of the Brahmaputra River flowing far below. The largest river in the Indian subcontinent in terms of flow, the Brahmaputra has begun far upstream in Tibet and has many more miles to go before it reaches the

W. A. T. Logan, *A Technological History of Cold-War India, 1947–1969*, Palgrave Studies in the History of Science and Technology,
https://doi.org/10.1007/978-3-030-78767-7_6

Bay of Bengal in Bangladesh. As the girders of the bridge whoosh past, fellow-passengers in your compartment toss coins out the window and stick their hands out into the breeze above the river. On the other side now, your train passes Kamakhya Hill, the site of a famous temple, before rolling into the Guwahati train station in the heart of town.

Your route from Bengal to Assam has taken you on a tour of many of the great transformations that have taken place in the Indian subcontinent since the mid-twentieth century. There was Partition, of course, which left Assam isolated from the rest of India, its great river draining into a foreign country. And there was modernization and its fellow-travelers industrialization, attempts at technological indigenization, and infrastructural development. All of these themes are present in your rail route to Assam, and specifically in the bridge over the Brahmaputra.

Constructed over a period of less than four years, from January 1959 to October 1962, the bridge was the first permanent crossing of India's last major unbridged river. Known simply as the Brahmaputra Bridge during construction, it was dubbed Saraighat Bridge by Prime Minister Nehru in 1963, after a seventeenth-century battle in which the Ahoms (Assamese) fought back the Mughal army's last attempted invasion of Assam.[1]

Double-decked Saraighat Bridge consists of ten main spans and two shore spans, all constructed of steel double Warren trusses[2] and spanning a total distance of 4,258 feet across the river. When first built, the bridge carried two meter-gauge rail lines on the lower deck and a paved motorway and pedestrian pathways on the upper deck. The project also included the construction of road and rail approaches to the bridge, the remodeling of the Gauhati (Guwahati) train station, and the construction of a new railway marshaling yard at Narangi in eastern Gauhati. The total

[1] Memo, "Brahmaputra Bridge Contract," NFR to Rail Secretary, March 9, 1965, file no. 62/W5/BR1/17/1-135, Ministry of Railways, NAI; "Brahmaputra Bridge Named Saraighat," *Assam Tribune* (*AT*), June 8, 1963; Maheswar Neog, *Lachit Barphukan: The Victor of the Battle of Saraighat* (New Delhi: Publications Division, 1983), 78–80; Edward Gait, *A History of Assam* (1906; repr. Delhi: Surjeet Publications, 2006), 160–61.

[2] In American English, a *truss* is a structure made of multiple individual beams, known as *girders*; trusses are capable of spanning openings and may be used in bridges or roofs. In Indian English, *girder* refers to a complete truss as well as its individual members. Except in quotations, this chapter uses the words *truss* and *girder* in their American senses.

project cost just over 100 million rupees, of which roughly half was spent on the bridge itself.[3]

The Brahmaputra Bridge project was built on behalf of the Northeast Frontier Railway (NFR), the easternmost subdivision of India's national rail network. B.C. Ganguli, who would have the honor of cutting the ribbon across the bridge when it opened in October 1962, served as Chief Engineer on the project. Throughout planning and construction, Ganguli and NFR were closely supervised by the Railway Board in New Delhi, represented by Karnail Singh, Member (Engineering). All major decisions regarding the bridge had to be cleared by Singh and the board prior to implementation. Two private Indian firms constructed the bridge for NFR. Hindustan Construction Co. (HCC) of Bombay dug the foundations in the riverbed and constructed the concrete piers, while Braithwaite, Burn and Jessop (BBJ) of Calcutta constructed the twelve steel spans that stretched from pier to pier across the Brahmaputra.

The Railway Board awarded the contract for the foundations and piers directly to HCC, without calling for open bidding. For the trusses, the Railway Board called for international tenders, but nevertheless selected BBJ over eight other firms. BBJ's bid used the least foreign exchange, but this was not why it was chosen. The open tender for trusses was a survey of contemporary bridge construction technology outside of India. The Railway Board determined that BBJ's methods were sufficiently up-to-date, and therefore it was not necessary to transfer new technology into India.

The Brahmaputra Bridge project was thus almost entirely domestic. BBJ subcontracted with a British firm to design and analyze the trusses, but the rest of the construction techniques were already well established in India by the late 1950s. Furthermore, all of the capital and labor, and almost all the materials, originated in India. Unlike the aircraft in Chapter 4 and the power stations in Chapters 7 through 9, the bridge technology in this chapter was never part of a concerted government indigenization program. The knowledge required for modern bridge-building had already reached India in the late colonial and very early independence periods, mainly through the efforts of private industry. In fact, this is one

[3] "Brahmaputra Conquered After Fifty Years' Planning," *Assam Tribune Magazine*, June 17, 1962; B.C. Ganguli, "The Northeast Frontier Railway in Service of Nation," *AT*, April 10, 1962.

of the few cases in this book in which indigenization in the nationalistic sense appears actually to have occurred.

There are two lessons here. The first is that private industry played an important role in the autarkical programs of Nehruvian India, even if the public sector was growing faster and receiving the most publicity at the time. The second is that indigenization (especially the narrow nationalistic kind) rarely lasts forever. Bridge-building in India was "indigenized" in the 1950s and 1960s, but not later. Although the first bridge over the Brahmaputra was almost entirely Indian, later bridges over the same river have had to import technology that was developed outside of India in more recent decades.

The Indian Railways and Assam

The modern Republic of India has one of the largest rail networks in the world. Trains are the main means of long-range travel in the country, and India's large, mobile population rides trains in huge numbers every day of the year. Reserved seats on important mainline trains can be wait-listed months in advance. The Northeast Frontier Railway, one of the seventeen railway zones in India, serves Assam with connections to all the neighboring states. As of 2016, NFR had 2,500 route-miles of track and carried 95 million passengers annually.[4]

The Indian rail network got its start during the colonial period, and British-era infrastructure is still very much in evidence throughout the Indian Railways. There are stately Gothic train stations, sturdy steel bridges, and more mundane structures like water tanks and maintenance sheds. But as you observed on your imaginary journey to Assam, the railways have grown and changed much since independence as well, with nationalization, reorganization, construction of new lines and stations, conversion from steam to diesel or electric locomotives, introduction of domestically produced locomotives and rolling stock, and conversion of lines to broad gauge.

The first rail line in India opened in 1853, and by 1920, India had a total of 38,500 miles of track, more than Germany and less than only the United States, Canada, and the Soviet Union. Among these countries,

[4] "Annual Performance Report 2015–16," *Northeast Frontier Railway*, http://www.nfr.indianrailways.gov.in/uploads/files/1491289045611-2015-16.pdf (accessed May 10, 2018).

India was the only one that was still a colony. It was not uncommon in the colonial world for railways to be built as single resource-extraction lines from mines or plantations to ports. In India, commercial competition and the political unification of the subcontinent under a single colonial administration meant that rail lines could, and did, develop into dense networks. The first Indian rail lines were built for the commercial interests of the colonizers, but the Indian public soon started riding trains in large numbers. It was thus from the transportation of India's huge population, rather than freight, that the railways became profitable.[5]

Daniel Headrick argues in *The Tentacles of Progress* that the colonial development of railways represented technology transfer in one sense but not others. Locomotives, rolling stock, rails, signaling systems, and so forth moved from Europe or North America to India, but the only technology transfer was physical. Knowledge was not transferred at the same time. The engineers who laid out the railways in British India and the princely states were British, as were the engineers who operated the locomotives.[6]

As we will find as we study the Brahmaputra Bridge, Headrick has failed to appreciate the extent of knowledge transfer into colonial India. Although the chief engineers who designed the rail lines were British to a man, junior engineers and assistant surveyors were often Indian. These men could gain experience while working for the British before setting out on their own. Rail lines, bridges, and tunnels were often built by contractors based in India. Some of these companies had British management, like Braithwaite, Burn and Jessop, while others, like Hindustan Construction Company, were Indian-owned. By independence, Indian knowledge of modern construction techniques was extensive and deep.

Furthermore, as Headrick acknowledges, the railways did establish workshops for maintenance of locomotives, which also doubled as locomotive manufacturing centers. Seven hundred locomotives were built in India before independence, of which 444 were made at Baroda and Central India's workshop in Ajmer and 217 at East India Railway's facility in Jamalpur. But these seven hundred locomotives represented

[5] Daniel R. Headrick, *The Tentacles of Progress: Technology Transfer in the Age of Imperialism, 1850–1940* (New York: Oxford University Press, 1988), 53–56, 60, 63; Arup Kumar Dutta, *Indian Railways: The Final Frontier—Genesis and Growth of the North-East Frontier Railway* (Guwahati: Northeast Frontier Railway, 2002); 9–11.

[6] Headrick, *Tentacles of Progress*, 81.

only 4 percent of the locomotives used in India before independence. By comparison, 14,420 locomotives (80 percent) came from Britain, and the balance from Germany or the United States. When it came to domestic production of locomotives, the limiting factor was not a lack of knowledge, but the resistance of the colonial government. The British government discouraged domestic production in India in order to protect British industry from colonial competition.[7]

By the beginning of the 1930s, railroads had been built across almost the entirety of the British Empire in India, from the southern tip of the subcontinent to Peshawar near the Khyber Pass. In the northeast, two separate lines ran into Assam. From Calcutta, a broad-gauge (5 ft 6 in) line of the government-operated Eastern Bengal Railway ran to Parbatipur in eastern Bengal, where it connected with a meter-gauge (3 ft $3\frac{3}{8}$ in) line that ran along the western and northern side of the Brahmaputra to Amingaon. The second route, belonging to the private Assam Bengal Railway and built entirely in meter gauge, began in Chittagong in eastern Bengal and ran through the Surma and Barak valleys to Upper Assam. In 1942, the colonial government nationalized the Assam Bengal Railway and merged it with the Eastern Bengal Railway.[8]

Until the second decade of the twentieth century, the railway business in India was booming. But then the two world wars and the global economic depression of the 1930s checked the growth of the railways. In World War I, the railways were forced to send locomotives, rolling stock, rails, and sleepers to Iraq. In the depression, the railways suffered a drop in revenue, and they were unable to keep up with maintenance. The already deteriorating railways were further affected when World War II forced them to surrender more equipment for military use. Trains and 4,000 miles of track were turned over to the military, plowshares were beaten into swords, and railway workshops were converted to armaments production.[9]

[7] Headrick, *Tentacles of Progress*, 81–82.

[8] Headrick, *Tentacles of Progress*, 67; Northeast Frontier Railway, "Brahmaputra Bridge Project (Amingaon): Project Report and Construction Estimate" 1959, file no. 59/W5/BR1/19/1-22, Ministry of Railways, NAI (hereafter cited as NFR, Brahmaputra Bridge Project Report [1959]).

[9] Planning Commission, *First Five Year Plan*, http://planningcommission.nic.in/plans/planrel/fiveyr/1st/welcome.html (accessed January 13, 2014); Headrick, *Tentacles of Progress*, 78–81.

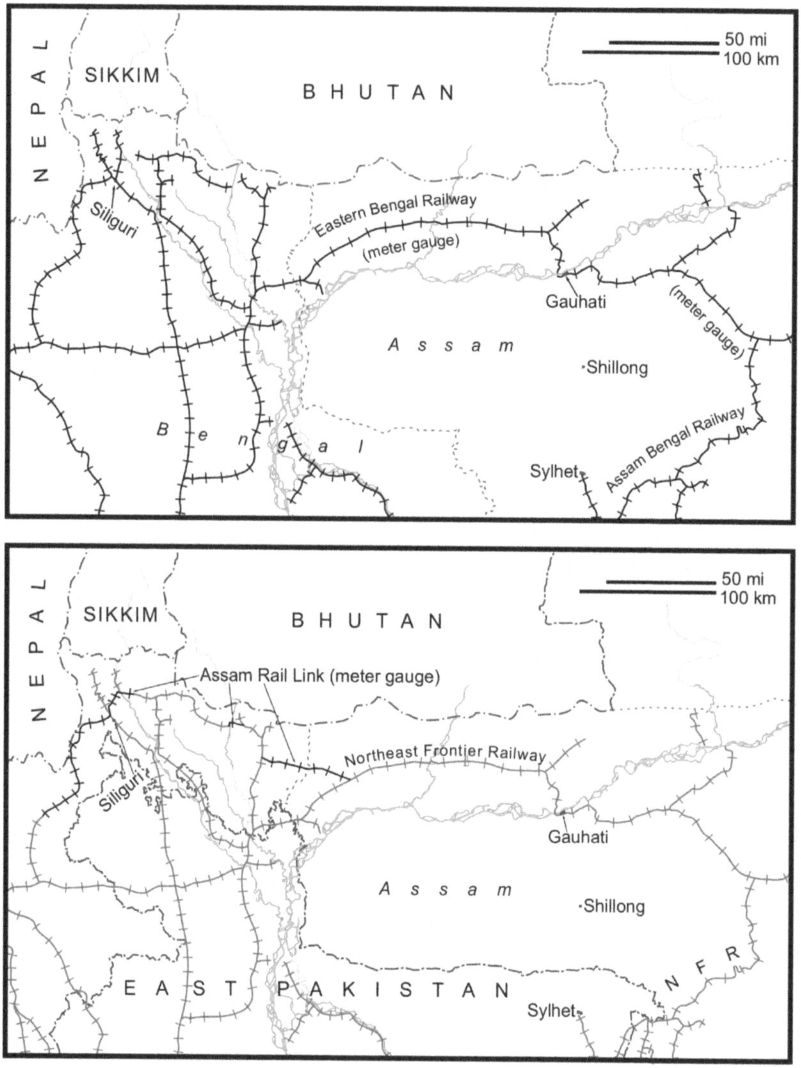

Fig. 6.1 Maps of rail lines in northeast India. Before Partition (*top*), the rail lines to Assam ran through eastern Bengal. After Partition (*bottom*), when eastern Bengal became East Pakistan, new lines had to be constructed to connect Assam with the rest of India through the Siliguri Corridor (Drawn by the author)

Two years after the war, the poorly maintained railways suffered another setback as new international borders cut rail lines in two. The western railways of India lost most of their workshops to Pakistan. In the Northeast, the two major pre-Partition rail lines into Assam had run through the gentler terrain of eastern Bengal, which was now East Pakistan (Fig. 6.1). Assam's existing transportation infrastructure was therefore cut off from the rest of India.

In response, the Railway Board launched a crash program in 1948 to build meter-gauge rail lines to connect Assam's orphaned rail network with the tracks of the rest of the country. In just under two years of work, engineers and laborers under the command of engineer-in-chief Karnail Singh completed the Assam Rail Link, which passed only through Indian territory. Freight trains first ran on the completed link in December 1949, with passenger trains beginning service the following month. Construction of the link included the laying of 142.5 miles of track in four sections and installation of twenty-two bridges with deep-well foundations, the longest of which was 1,425 feet in length.[10]

Partition also brought immediate changes to Assam, the territory isolated from the rest of India by East Pakistan. An influx of displaced persons at and after Partition contributed to a dramatic rise in the population of Gauhati, Assam's major urban center. Although the town had largely been bypassed by pre-independence industrialization, Gauhati had long been the commercial and transportation hub of Assam. The city's name, which derives from Assamese words meaning "areca nut market," reflects the city's historic status as an economic center. (The current spelling of the name is "Guwahati," which more closely represents the pronunciation of the city's name in Assamese.) In the 1901 census, the city registered a population of 11,661. Fifty years later, Gauhati had grown to the modest size of 43,615. Then by the 1961 census, the city's population had exploded to 100,707[11] (Fig. 6.2).

[10] "Assam Rail Link Project" (n.d. [1950?]), in "Railroads in India, Part 1," *South Asia Ephemera Collection* (New Delhi: Library of Congress Office, 1994); Ratan Lall, "N.F. Railway Does It," *Assam Tribune Magazine*, April 16, 1961; D.C. Baijal, "The North-East Frontier Railway," *Overseas Railways*, 1958, 51–52.

[11] "An Introduction to Tentative Master Plan for Greater Gauhati," *Assam Tribune* Greater Gauhati supplement, November 25, 1958; E.H. Pakyntein, *Kamrup District Census Handbook, Census of India 1961* (Government of Assam, 1964), Assam State Archives.

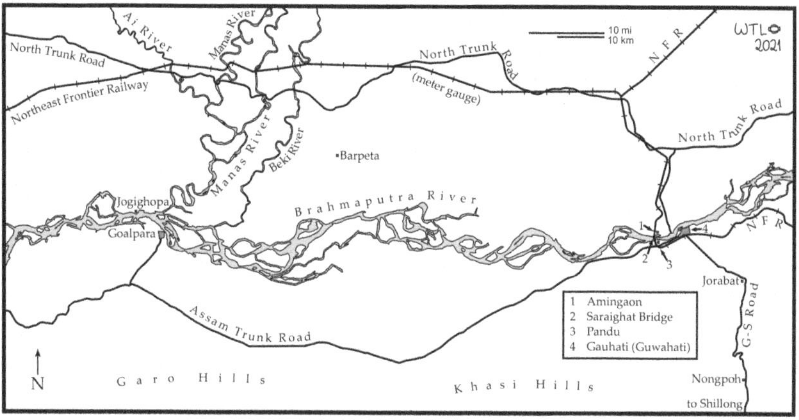

Fig. 6.2 Map of the Lower Brahmaputra in the 1960s (Map drawn by the author, based on US Army Map Service maps at http://www.lib.utexas.edu/maps/ams/india/, accessed April 21, 2016)

Government planning brought changes to Assam as well, even if the plans never fully achieved their ambitions. Development was organized under both state-level and central government-level five-year plans. Under the First Five-Year Plan of the Union (central) government (1951–1956), hundreds of miles of flood-control dikes were built alongside the Brahmaputra. During the second and third plans, new electric power projects made power available for industrial development. (The biggest of these projects, Umiam Hydroelectric in the Khasi Hills, is the subject of Chapter 7.) The government also established public-sector industrial operations to process Assamese resources in Assam. One of these was Noonamati Refinery in Guwahati, completed in 1962 with Romanian technical collaboration. A sixteen-inch pipeline connected the refinery with the Nahorkatiya and Moran oilfields in upper Assam.[12]

[12] E.H. Pakyntein, Kamrup District Census Handbook, Census of India 1961 (Government of Assam, 1964), Assam State Archives; "Nehru Inaugurates Gauhati Refinery This Morning," *AT*, January 1, 1962; "River Crossing Every Nine Miles for Oil India Pipeline," *International Oilman*, December 1959, 407–8. Arupjyoti Saikia argues that the construction of embankments was politically motivated, as plantations of tea, jute, and other cash crops received more attention than plots of subsistence agriculture. "Jute in the Brahmaputra Valley: The Making of Flood Control in Twentieth-Century Assam," *Modern Asian Studies* 49, no. 5 (September 2015): 1433–34.

Development of transportation infrastructure also received attention in the Five-Year Plans of the Union and state governments. Under the direction of the state plans, the Assam Public Works Department expanded roads, repaired damage, and replaced mile or furlong posts along the highways with kilometer posts. Two national highways ran along opposite sides of the Brahmaputra: the Assam Trunk Road, which passed through Gauhati, and the North Trunk Road. Before independence, both had been largely unpaved, and the North Trunk Road was passable only in the dry season. By the mid-1960s, the Assam Trunk Road was partially paved.[13]

In the plans of the early independence period, roads were secondary and railways were supreme. In Assam, the plans of the fifties and sixties focused on improving the Assam Rail Link. As constructed just after independence, the Link provided permanent connections only to the right (north) bank of the Brahmaputra. At Amingaon, trains transferred to a ferry, which crossed the river to Pandu, five miles downstream of Gauhati. The ferry was inefficient and slow, with as much as a six-hour delay as carriages were uncoupled and individually shunted onto the ferry. The maximum capacity of the ferry was only 250 carriages a day. The ferry could be dangerous, and service often had to be canceled during the seasonal monsoon floods. In addition to the railway ferry, smaller ferries operated by the state Public Works Department crossed the river at various points. One of the ferries sailed regularly from Gauhati proper to North Gauhati, a suburb across the river. This ferry could also be dangerous. In February 1958, the ferry sank after getting caught in a sudden hailstorm, and eight of the twenty people on board drowned.[14]

[13] Shyam Bhadra Medhi, *Transport System and Economic Development in Assam* (Gauhati: Assam Publication Board, 1978), 41; Priyam Goswami, *The History of Assam from Yandabo to Partition, 1826–1947* (Hyderabad: Orient Blackswan, 2021), 187–88; Government of Assam Public Works Department (PWD), "Report on the Administration of Public Works Department (Roads & Buildings Wing), Assam for the Year 1961–62," Assam State Archives; Government of Assam PWD, "Report on the Administration of Public Works Department (Roads & Buildings Wing), Assam for the Year 1963–64," Assam State Archives; Bipin Borgohain, "Calcutta to Gauhati by Road," *AT*, January 3, 1960.

[14] NFR, Brahmaputra Bridge Project Report (1959); Dutta, *Indian Railways: The Final Frontier*, 278; Northeast Frontier Railway, "50 Glorious Years of Saraighat Bridge" (Guwahati, 2012), author's collection; "Boat Disaster in the Brahmaputra Between Gauhati and North Gauhati," *Proceedings of the Assam Legislative Assembly*, May 2, 1958, in *The Assam Gazette* Part VI, January 20, 1960, 3477–80.

As state and central authorities were developing the transportation infrastructure of Assam, the Ministry of Railways split all of Assam and parts of West Bengal and Bihar off of the North East Railway to form the Northeast Frontier Railway (NFR). On January 15, 1958, at a ceremony held at Pandu, the headquarters of the new railway zone, Railway Minister Jagjivan Ram flipped a switch to cause a map lit in a single color to change to two different colors, one for the North East Railway and the other for the new Northeast Frontier Railway. It would be this new railway zone that would undertake the major project of constructing the Brahmaputra Bridge.[15]

Prehistory of the Brahmaputra Bridge

The Northeast Frontier Railway was the latest in a series of institutions that attempted to bridge the Brahmaputra River. Even in the precolonial period, there is a historical account of an eighteenth-century Ahom general, Kalia Bhomara, collecting timbers in preparation for bridging the river near Tezpur. There is no evidence that any part of this bridge was ever constructed, and indeed it is unlikely that such a span could have been made with eighteenth-century Assamese technology. The Ahoms did have the ability to build permanent masonry bridges, using technology imported from Bengal. One of these bridges, the 200-foot Namdang Stone Bridge near Sibsagar, is still used by a modern highway. But spanning a river as large and capricious as the Brahmaputra was not feasible without industrial technologies like steel and concrete.[16]

In 1910, the Eastern Bengal Railway made the first modern plan to construct a permanent bridge over the Brahmaputra. The railway dropped the plan without implementing it, then resumed work on it in the 1930s. With the outbreak of World War II, the colonial government began to pay serious attention to the problem of bridging the river. After the Japanese conquered Burma, northeast India became a theater of war. The Japanese crossed the Indian border and captured parts of the princely state of Manipur. To facilitate the transport of materiel to the front, the colonial

[15] "New Railway Zone Inaugurated," *AT*, January 16, 1958.

[16] S. Ponnuswamy and V.C. Sharma, eds., *Bridging River Brahmaputra: Past Present and Future* (Guwahati: Northeast Frontier Railway, 1988), 17; Subhra Das and S.H. Pargiri, "Namdang Stone Bridge," Indira Gandhi National Centre for the Arts, http://www.ignca.nic.in/asi_reports/ASSSIV_015.pdf (accessed October 30, 2015).

government built bridges across certain strategic waterways. (One of these was the iconic Howrah Bridge in Calcutta.) After preliminary studies of the construction of a bridge across the Brahmaputra, the colonial government decided that it would be more practical to expand the existing Amingaon–Pandu ferry instead.[17]

Proposals to bridge the Brahmaputra remained relevant after the end of the war and Partition, as the population growth of Assam was putting increased strain on existing transportation infrastructure. During the latter half of the 1940s and into the 1950s, the siting of the bridge remained an open question. Not all parts of the river were equally bridgeable from geographical, engineering, and economic standpoints. Only two sites received serious consideration, Jogighopa and Gauhati, as these are the only two places in lower Assam where the river's channel is constrained by rock outcroppings. In 1946 and 1948, the Bengal and Assam Railway dispatched surveyors to Jogighopa, seventy miles downstream from Gauhati. Using water borings and electrical resistivity surveying (in which electric current was run between electrodes spaced at different points in the river bed), the surveyors determined that the depth of bedrock varied widely across the proposed route of the bridge, ranging from more than 350 feet near the north bank to only 100 feet near the south bank. Engineers for the railway deemed it unwise to build some piers on bedrock and other piers on sediment, because they wanted the piers to respond to earthquake stresses uniformly. For this reason, they recommended against siting the bridge at Jogighopa.[18]

Economic and geographical considerations also favored Gauhati over Jogighopa. At Gauhati, the river is not as wide as it is at Jogighopa. The narrowest part of the river, 2,600 feet at Kamakhya Hill, was unsuitable because the water was swift and turbulent at that point. Further downstream, the channel widened and the river slowed. Between Pandu and Amingaon, the river was 4,000 feet wide, which was still less than half the width of the 9,600-foot channel at Jogighopa. A narrower river channel

[17] NFR, Brahmaputra Bridge Project Report (1959); B.C. Ganguli, note on Brahmaputra Bridge for the Railway Board, September 7, 1957, file no. 57/W5/BRI/17 (Contract)/1-135, Ministry of Railways, NAI; S. Ponnuswamy, "Challenges in Bridging River Brahmaputra," *Indian Concrete Journal* 71, no. 8 (August 1997): 399.

[18] "Bridge Site: Jogighopa, Goalpara District, Assam," *Records of the Geological Survey of India* 79 (1954): 820; L.N. Kailasam, "Some Results of the Application of Resistivity Methods to Problems of Civil Engineering, Mining and Ground Water in India," *Bulletins of the Geological Survey of India*, Series B: Engineering Geology and Ground-Water, no. 3 (Calcutta: Government of India Press, 1952).

meant fewer piers and fewer spans, and therefore a cheaper bridge. At the Gauhati site, neither bank of the river channel needed stabilizing, while the south bank at Jogighopa would need an embankment. While the railway already crossed the river by means of the ferry at Gauhati, a bridge at Jogighopa would need an additional twenty miles of track from Bongaigaon on the north bank and sixty miles of track to Goalpara on the south bank.[19]

There was one serious disadvantage with Gauhati, which threatened to negate the benefits of the site. The rail line to Amingaon crossed the basin of the Manas River, which joins the Brahmaputra just upstream from Jogighopa. The Manas is one of several rivers that flows down from the Bhutanese Himalaya between Jogighopa and Amingaon. The geology of the north-bank tributaries of the Brahmaputra, which have steep slopes, sandy beds, and high silt loads, makes them susceptible to flooding or shifting into new channels. By contrast, the south-bank tributaries, flowing out of the Garo and Khasi Hills of the Meghalaya Plateau, have flatter grades, and they meander from the foot of the hills through the floodplain to the Brahmaputra. Therefore, they are much less prone to flooding.[20]

In 1942, serious flooding in the Manas basin caused damage to the rail line and made the railway authorities consider circumventing the troublesome tributary by bridging the Brahmaputra at Jogighopa. The Bengal and Assam Railway formed a study committee in 1944 to consider the problem. The committee finally delivered its report in 1948, stating that geological concerns clearly favored avoiding the Manas basin. But by this point Partition had added an additional political dimension to the question of siting a bridge. Although the 1948 report initially seemed to favor the Jogighopa site, it made an about-face in its conclusion. Citing the "closeness of the boundary of Pakistan," and the likelihood that a bridge at Jogighopa would carry less traffic than one near Gauhati, the report recommended that "the proposed bridge at Jogigopa [*sic*] may be deferred and one with far less cost may be built at Amingaon." Later, the Assam Link Stabilization Committee determined that the line through the Manas basin could be stabilized. Thus, for geological, geographical, economic, and political reasons, Gauhati emerged as the clearly favored

[19] NFR, Brahmaputra Bridge Project Report (1959).

[20] Ibid.

site for a permanent bridge across the Brahmaputra. In communications later in the 1950s, members of the Railway Board and NFR repeatedly cited these arguments in favor of a bridge near Gauhati rather than Jogighopa.[21]

In 1956, the Second Five-Year Plan made reference to the railways' intention to begin construction of "three important bridges" during the plan period, "one each across the Brahmaputra, the Jumna and the Gandak" On September 2, 1957, Karnail Singh called a meeting in his office in the North Block secretariat, New Delhi, to initiate preparations for the Brahmaputra Bridge. A former engineer for the North Western Railway and chief engineer of the Assam Rail Link, Singh had recently been appointed Member (Engineer) of the Railway Board. At the first Brahmaputra Bridge meeting, Singh identified two matters that needed immediate attention: finalizing the location of the bridge and selecting the contractor to construct the piers. It was imperative to begin preparations for constructing the bridge's foundations as soon as possible, because the five-month working season in Assam would come to an end when the *chhota barsaat* (pre-monsoon storms) began in March. "The matter is most urgent" Singh wrote in a memo about the meeting, "and any delay, of a few days or weeks now, would mean a year's delay in execution of works in the field."[22]

By this point, B.C. Ganguli had already been selected as Chief Engineer of the Brahmaputra Bridge project for the North Eastern Railway. On September 7, 1957, Ganguli wrote a note to the members of the Railway Board requesting that they approve the site of the bridge as Amingaon–Pandu, near Gauhati, rather than Jogighopa. The Railway Board gave its approval. In November, Minister of Railways Jagjivan Ram

[21] Ibid.; "Summary Record of the Meeting Held in the Room of the Member, Engineering, Railway Board on 2nd September, 1957," file no. 57/W5/BRI/17 (Contract)/1-135, Ministry of Railways, NAI; B.C. Ganguli, note on Brahmaputra Bridge for the Railway Board, September 7, 1957, ibid.; [Jagjivan Ram] to Jawaharlal Nehru, November 1957, ibid.

[22] Planning Commission, *Second Five Year Plan*, http://planningcommission.nic.in/plans/planrel/fiveyr/2nd/welcome.html (accessed January 13, 2014); "Summary Record of the Meeting Held in the Room of the Member, Engineering, Railway Board on 2nd September, 1957," file no. 57/W5/BRI/17 (Contract)/1-135, Ministry of Railways, NAI; Karnail Singh, "Improvements in Rail Communications Between Assam and the Rest of India," memo, September 5, 1957, ibid.; Ranjit Singh, ed., *Sikh Achievers* (New Delhi: Hemkunt Publishers, 2008), 98–99.

wrote to Jawaharlal Nehru, informing him of the plans to construct the bridge. In reply, the Prime Minister indicated his agreement with the plans and asked that Ram announce the decision in Cabinet, which he did on November 20.[23]

When it came to the question of which contractor should build the foundations of the bridge, the Railway Board and NFR never seriously considered any firm other than Hindustan Construction Company (HCC). HCC had been founded in 1926 by Seth Walchand Hirachand (1882–1953), a Maratha Jain businessman. Like many Indian industrialists of his time, Hirachand had diverse interests, and he would also establish the Scindia Steamship Navigation Company, Hindustan Shipyard, and Hindustan Aircraft (discussed in Chapter 4). Hindustan Construction Company's first contract was for a railway tunnel in the Bhor Ghat above Bombay. In the following decades, the company built up a diverse portfolio of projects executed for the railways, state public works departments, municipalities, and private industry. Just after independence, HCC received a contract to build bridges for the Assam Rail Link. The company's subsequent experience on a bridge over the Ganga (Ganges) at Mokameh Ghat, Bihar convinced the railways that the company was qualified to build the Brahmaputra Bridge.[24]

But before the railways could open negotiations with HCC, an offer to build the piers of the Brahmaputra Bridge came out of nowhere from another company. The Railway Board received a letter from M.R. Venkataram, who was writing on behalf of Hind Constructions Ltd., which was headquartered in either Madras or Calcutta (the letter left this point unclear). "I wish to lay before you," Venkataram wrote, "that the Hind Constructions Ltd., can certainly undertake Bridges of that

[23] B.C. Ganguli, note on Brahmaputra Bridge for the Railway Board, September 7, 1957, file no. 57/W5/BRI/17 (Contract)/1-135, Ministry of Railways, NAI; H.D. Awasty, memo, September 14, 1957, ibid.; [Jagjivan Ram] to Jawaharlal Nehru, November 1957, ibid.; Jawaharlal Nehru to Jagjivan Ram, November 18, 1957, ibid.

[24] Gita Piramal, *Business Legends* (New Delhi: Penguin Books, 1998), 160–68, 220–28; HCC, "Statement of Works Executed and in Hand for Year Ending 31-12-48," S. No. 341, HCC subject files, papers of Walchand Hirachand, NMML; HCC, "Directors' Report and Statement of Accounts for the Year Ended 31st August 1954," S. No. 342, ibid; Ganga Bridge Construction Co. Ltd, "Tender for the Construction of Main Bridge Over River Ganga Near Mokameh (Excluding Training and Approach Works)," May 28, 1954, file no. 62/W5/BR1/17/1-135, Ministry of Railways, NAI; H.K.L. Sethi, "Bridge Across the Ganga Near Mokameh," March 13, 1959, file no. 59/W5/BRI/6/1-48, Ministry of Railways, NAI.

magnitude with confidence. We have about 3 air-locks two of which are in commission at present on the Highway Bridge at Mandvi, Surat." Venkataram went on to explain that the company also had a contract on a bridge near Kharagpur (West Bengal), and had also "taken two crores [20 million rupees] of work at the Roorkela Steel Plant." Venkataram guaranteed the Railway Board that his company was fully qualified to work on a project as big as the Brahmaputra Bridge.[25]

After receiving this surprising request, Railway Secretary R.E. de Sa wrote some letters to check the credentials of Hind Constructions, and in the process found that the firm was something less than it claimed to be. S.K. Mukerjei, Deputy Secretary in the Ministry of Defence, reported on Hind Constructions' poor performance on the Naval Dockyard Expansion Scheme. The firm had been contracted to dredge, reclaim land, and construct wharves, but its progress was "extremely slow" and it ultimately abandoned the work and forfeited the contract. In another response to de Sa's background-checking, D.N. Chopra reported that Hind Construction was not employed as contractor or subcontractor at Rourkela Steel Plant. A third response, anonymous and undated, gave a further indictment of Hind Constructions' already poor credentials. The firm had submitted a tender to build a combination road and rail bridge over the Godavari River near Rajahmundry. "When the tenders were opened," the report states, "the Company withdrew their tend as their tenders were so ridiculously low."[26]

With Hind Constructions eliminated from consideration, the Railway Board opened negotiations with HCC. On January 2, 1958, S.S. Godbole, Deputy General Manager of HCC, wrote a letter to the Railway Board laying out his firm's offer for the Brahmaputra Bridge. Godbole argued that certain factors would make a bridge over the Brahmaputra more expensive than the recently completed Ganga Bridge, including recent increases in the prices of cement, steel, and coal, and the need to recruit labor and staff from out of state. For these reasons, Godbole offered to build the bridge at a 7.5 percent higher rate than the Ganga Bridge. But just two weeks later, Godbole sent a second letter to the

[25] M.R. Venkataram to N.K. Roy, October 2, 1957, file no. 57/W5/BRI/17 (Contract)/1-135, Ministry of Railways, NAI.

[26] R.E. de Sa to R.E. Aserappa, October 25, 1957, file no. 57/W5/BRI/17 (Contract)/1-135, Ministry of Railways, NAI; S.K. Mukerjei to R.E. de Sa, November 8, 1957, ibid.; D.N. Chopra to Karnail Singh, November 11, 1957, ibid.; Memo, n.a., n.d., ibid.

Railway Board, backing down from his request for higher rates. This time, he offered to do the work for 2.64 percent below the Ganga Bridge rates, provided the Railway Board offer for certain concessions, including the use of a barge from the rail ferry for a floating crane and land near the construction site for camps, offices, workshops, and other project needs. The Railway Board found these terms acceptable, and on January 18, the Rail Secretary authorized HCC to start preparations for construction immediately.[27]

The Search for a Second Contractor

With the selection of Hindustan Construction Company to build the piers, NFR had only half the contractors it needed to build the Brahmaputra Bridge. Another company would have to be selected to build the trusses that would reach across the river on HCC's piers. An obvious choice for the second contractor was the Calcutta-based firm Braithwaite, Burn and Jessop, which had built the Mokameh Ghat bridge in partnership with HCC. But rather than award the contract directly to BBJ, NFR, and the Railway Board called for open bidding on the project—only to award the contract to BBJ in the end. BBJ's receiving the contract may have been a foregone conclusion. What is important for our study is not so much the outcome as the process, and the motives underlying the call for bids.

The favorite throughout the tender process, Braithwaite, Burn and Jessop was founded in 1935 as a consortium of three Calcutta-based engineering companies. The consortium built the iconic Howrah Bridge in its hometown and went on to execute contracts for the railways after independence. Unlike HCC, the constituent companies of BBJ were founded by British, not Indian, capitalists, and the consortium remained under British management into the 1950s and 1960s. The founders were not nationalists and would never be given the glowing tributes offered to the likes of Hirachand and the Tatas and Birlas. But since BBJ was based in Calcutta and accepted payment in rupees, it qualified as an Indian company.[28]

[27] S.S. Godbole to Rail Secretary, January 2, 1958, file no. 62/W5/BR1/17/1-135, Ministry of Railways, NAI; Rail Secretary to HCC, January 18, 1959, ibid.

[28] Proceedings of the tender committee for Brahmaputra Bridge girders, August 20, 1959, file no. 58/W-5/BRI/7/1-132, Ministry of Railways, NAI; "Company Profile /

After deciding to call for open bids, NFR wrote up a tender notice and detailed tender documents for the design, fabrication, erection, and maintenance for one year (from date of completion) of ten spans of 397 feet each, plus two shore spans of 104 feet, 9 inches. The dimensions of the proposed trusses matched the piers that HCC was preparing to build. NFR sent the draft tender documents to the Railway Board for approval. The Railway Board returned the draft with orders to make two changes. First, the railway deck of the trusses, proposed to be built with two meter-gauge rail lines, should be readily convertible to carry a single broad-gauge line. Second, an engineer of the railways should be seconded to the contractor and "placed in charge of the erection of girders [trusses]." The intention of this clause was for the railways to have control over construction and also learn from the contractor's experience.[29]

On August 9, B.K. Mitra of NFR delivered 120 copies of the tender notice to the Railway Board, half of which were earmarked for delivery to the diplomatic missions of foreign countries in New Delhi. The Soviet Embassy initially expressed interest in the "Braxmanympa Bridge" tender, but later Technoexport wrote from Moscow with regrets that it was already overcommitted and did not intend to tender for the bridge. By March 31, 1959, the date of the opening of the tenders, NFR had received tenders from nine firms, based in India, West Germany, France, Yugoslavia, and Japan. With three firms submitting multiple options, NFR received a total of eighteen tenders.[30]

On June 10, 1959, NFR formed a tender inspection committee consisting of two engineers (one of whom was B.C. Ganguli) and an accounts officer, G.F. Penn Anthony. Members of the Railway Board also contributed to the inspection of the tenders by cross-examining the representatives of the tendering companies. The NFR committee quickly determined that twelve of the tenders had inherent flaws and could be

About Us," *Bharat Bhari Udyog Nigam Ltd.*, http://bbunl.com/profile.html (accessed November 9, 2015). The Indian government nationalized BBJ in 1987, and in July 2015 the company was amalgamated into Bharat Bhari Udyog Nigam Ltd. (India Heavy Industry Corporation Ltd.) or BBUNL.

[29] M.N. Bery to B.C. Ganguli, March 7, 1958, file no. 58/W-5/BRI/7/1-132, Ministry of Railways, NAI; M.N. Bery to NFR General Manager, July 26, 1958, ibid.; NFR, "Tender Documents for the Brahmaputra Bridge at Amingaon," ibid.

[30] V. Sergeev to Rail Secretary, September 30, 1958, file no. 58/W-5/BRI/7/1-132, Ministry of Railways, NAI.; N. Trufanov to Rail Secretary, January 10, 1959, ibid.; G.P. Warrier, memo, September 22, 1959, ibid.

eliminated from consideration immediately. The six remaining tenders represented just three companies: three tenders from Ingra Zagreb (Yugoslavia), two from Gollnow-Werke (Germany), and one from BBJ.[31]

The committee studied the six remaining tenders in detail, working through the calculations, and determining where the structures would need to be strengthened. Because of design modifications and other considerations such as customs duty and supply of facilities by the railway, the committee adjusted the tender estimates. Before adjustment, the lowest bid was 21.2 million rupees (Rs.) for rhomboidal trusses from Ingra Zagreb. Gollnow-Werke's lowest estimate was Rs. 21.8 million, and BBJ's single estimate was Rs. 22.7 million. Since both Ingra Zagreb's and Gollnow-Werke's designs needed strengthening, the tender committee increased the estimates. At the same time, the committee found BBJ's design suitably strong and even subtracted Rs. 179,000 from the estimate. With these adjustments, BBJ's tender became the lowest by a narrow margin of Rs. 12,000.[32]

The total cost of the tender was not the only financial consideration; the foreign-exchange component was also a central factor in the tender committee's deliberations. In this regard, BBJ enjoyed an advantage from the beginning, because the Indian firm's tender called for Rs. 7.4 million to be spent abroad, mainly for importing British steel. By comparison, Gollnow-Werke's two tenders each called for Rs. 8 million of foreign exchange, and all of Ingra Zagreb's alternatives required Rs. 9.3 million.[33]

The tender committee also expressed a preference for BBJ because the firm was Indian and had already worked with the country's railways. The report of the committee's August 20 meeting stated that of the companies that tendered bids, only Ingra Zagreb, Gollnow-Werke, and BBJ were really qualified to build the bridge. Ingra Zagreb had recently carried out some contracts for the Indian Railways and had also built major bridges in Yugoslavia. On the basis of this prior experience, the committee concluded that Ingra Zagreb's ability to build the Brahmaputra Bridge's

[31] G.P. Warrier, memo, September 22, 1959, file no. 58/W-5/BRI/7/1-132, Ministry of Railways, NAI; Proceedings of the tender committee for Brahmaputra Bridge girders, August 20, 1959, ibid.

[32] Ibid.

[33] Proceedings of the tender committee for Brahmaputra Bridge girders, August 20, 1959, file no. 58/W-5/BRI/7/1-132, Ministry of Railways, NAI.

superstructure "is, therefore, not denied." Regarding Gollnow-Werke's qualifications, though, the committee had some doubts. "M/s. Gollnow Werke are understood to have carried out works of fabrication and erection of big girders [trusses]," the committee stated blandly, "and they should be able to successfully carry out this contract." In the copy of the report in the Railway Ministry's file, the word "should" is struck out, and "may" written above it.[34]

Braithwaite, Burn and Jessop, by comparison, had impeccable qualifications. The committee cited the company's experience with the Howrah Bridge, Chakki Bridge, and Malviya Bridge, as well as the recent Rajendra Bridge (Ganga Bridge). Because of these qualifications, "they are fully capable of undertaking this work." Since BBJ had the lowest estimate (after adjustments) and the lowest foreign-exchange cost, the tender committee recommended that it be selected as the contractor for the superstructure. Furthermore, since this was a high-profile project, it would "definitely add to the prestige and experience of the firm undertaking the work," and it was preferable that an Indian firm gain this prestige—even if its owners were British expats. Additionally, the committee mentioned the potential that foreign aid on a project could be cut off if international politics changed. It was thus safer to entrust the work to an Indian company. Therefore, the decision to select BBJ was influenced by considerations of national pride and autarky as well as overall costs and foreign exchange.[35]

The Railway Board deferred a final decision until the tendering companies could provide further information about their proposals. On October 20, representatives of the three eligible companies met with Karnail Singh and other members of the Railway Board and NFR to discuss the tenders. The committee told Gollnow-Werke's representative, Serge de Pahlen, that the firm had not provided stress sheets and influence line diagrams, and without these, their design was difficult to evaluate. Ingra Zagreb's representatives learned that their proposal also had shortcomings, as it omitted a complete erection plan, with calculations of stresses that would develop in the girder members at each step. BBJ's representatives were told that their design was satisfactory.[36]

[34] Ibid.

[35] Ibid.

[36] Memo regarding Brahmaputra Bridge project meeting held on October 20, 1959 in North Block, file no. 58/W-5/BRI/7/1-132, Ministry of Railways, NAI.

Ingra Zagreb and Gollnow-Werke provided further clarifications to their designs and proposed incentives to make their offers more appealing. BBJ also proposed an alternative to its earlier tender. Rather than importing British steel, as the company had done on the Ganga Bridge project, BBJ could use Indian steel as far as possible. Even though it was a domestic product, Indian steel was actually more expensive than imported steel at this time, because India's steel mills had a limited production capacity. The use of Indian steel would raise the price of the tender by Rs. 550,000, but it would also cause the foreign-exchange component to drop from Rs. 7.4 million down to only Rs. 600,000. The remaining foreign exchange would mostly pay for specialized steel components that could not be produced economically in India. The Railway Board found BBJ's offer to use domestic steel persuasive, and therefore it decided to accept the firm's offer, even though the added price of Indian steel would raise the cost of the bridge above some of the tenders for foreign firms. In this case, saving foreign exchange and promoting Indian industry proved to be more important considerations than spending the least money on the bridge overall.[37]

Not everyone was convinced. Serge de Pahlen, the representative of Gollnow-Werke, continued to argue his company's case long after the railways had decided in favor of BBJ. Count de Pahlen, as he styled himself in his letters, had arrived in India in August 1959 and settled into the Ashoka Hotel in the Chanakyapuri diplomatic enclave of New Delhi. From his room in the Ashoka, he addressed voluminous correspondence with government officials on hotel stationery. In response to the requests of the committee that met in Karnail Singh's office on October 20, de Pahlen forwarded further technical details for his company's design. He also offered to substitute the foreign-exchange requirements for the bridge with a barter deal, in which the export of Indian iron ore to Germany would take the place of hard currency. Even after the Railway Board accepted BBJ's offer on November 19, de Pahlen kept writing to

[37] G.P. Warrier, memo, September 22, 1959, file no. 58/W-5/BRI/7/1-132, Ministry of Railways, NAI; B.C. Ganguli, Note on tenders received for Brahmaputra Bridge project, attachment to letter to D.R. Kohli, July 3, 1959, ibid.; Citation: G.P. Warrier to NFR General Manager, November 19, 1959, ibid.

the board, offering to collaborate with Indian industry. When de Pahlen heard that BBJ's offer had been higher than Gollnow-Werke's, he wrote angry letters to the Railway Board, and even brought his case to the Finance Minister, Morarji Desai.[38]

By this point, the Railway Board had started filing de Pahlen's letters unanswered. When the Finance Ministry looked into the case, J. Dayal, financial adviser of the Railway Board, prepared a memo explaining that BBJ's offer had been lower than Gollnow-Werke's. "From the papers which I have examined and which are placed below," Dayal wrote, "it seems to me that the award of the contract to Messrs. B.B.J was done in a regular manner, and in the best interests, financial and otherwise, of the Government." The "otherwise" interests, left implicit in the letter, were promoting Indian industry and letting an Indian company gain the prestige of completing a large project like the Brahmaputra Bridge.[39]

The railways had originally planned to offer the contract for the superstructure to BBJ, and then after a lengthy tender process, BBJ ended up receiving the contract. Why, then, did the railways go to the trouble of calling for an open tender in the first place? B.C. Ganguli explained the rationale in a letter to D.R. Kohli of the Railway Board: "Global Tenders were called for this work with a view to collect up-to-date knowledge of the technological advancement in this type of work if any, and a financial advantage therefrom, if possible, commensurate with the expenditure of foreign currency." In calling for tenders, the railways were investigating the possibility of technology transfer. Modern bridge-building techniques were well established in India by this point, but the railways wanted to know if the technology had changed outside India, and if so, whether their own country would benefit from importing this technology.[40]

One way to study the state of foreign technology would have been for the railways to send a team of engineers on a study tour of current bridge

[38] Anang Pal to Home Secretary, February 6, 1960, file no. 58/W-5/BRI/7/1-132, Ministry of Railways, NAI.; S. de Pahlen to J. Dayal, October 22, 1959, ibid.; S. de Pahlen to J. Dayal, November 29, 1959, ibid.; S. de Pahlen to G.T. Venugopal, December 30, 1959, ibid.

[39] J. Dayal, memo, December 18, 1959, file no. 58/W-5/BRI/7/1-132, Ministry of Railways, NAI.

[40] B.C. Ganguli, Note on tenders received for Brahmaputra Bridge project, attachment to letter to D.R. Kohli, July 3, 1959, file no. 58/W-5/BRI/7/1-132, Ministry of Railways, NAI.

projects in industrialized nations. This had been done for other Nehruvian development projects. To cite one example, the project management of Bhakra Dam sent Jagman Singh, one of the junior engineers on the project, to the United States in 1952 to tour current construction projects including Folsom Dam near Sacramento and Detroit Dam in Oregon. Another strategy, adopted by NFR for the Brahmaputra Bridge, was to have the foreign countries come to them, and with their detailed tenders display the current state of their technology. From analysis of the tenders from foreign firms, NFR determined that Indian bridge-building techniques were suitably up-to-date, and there would be no benefit in employing a foreign firm.[41]

This did not mean that BBJ's design for the bridge spans was entirely indigenous. BBJ contracted Freeman Fox and Partners of London to design and analyze the bridge structure. Freeman Fox had also subcontracted the similar truss design for the Ganga Bridge. As per the conditions laid out in the tender documents, NFR engineer N.J. Vacha was seconded to BBJ in February 1960. In this capacity, he oversaw the design and analysis of the bridge spans in London. By doing so, Vacha gained experience that he brought back to the Indian Railways after completing his duties with BBJ. In this respect, technology transfer did take place in the Brahmaputra Bridge project, even though an Indian firm served as the prime contractor for the spans.[42]

Building the Brahmaputra Bridge

Hindustan Construction Company began building the Brahmaputra Bridge's piers in the dry season of 1958–1959, when the deepest part of the riverbed was only forty-five feet underwater. Construction of the

[41] Jagman Singh, *My Tryst with the Projects Bhakra and Beas* (New Delhi: Uppal Publishing House, 1998), 139–41.

[42] "Minutes of the meeting held in the office of the Engineer-in-Chief, Brahmaputra Bridge Project, Pandu on 22nd December, 1959," file no. 58/W-5/BRI/7/1-132, Ministry of Railways, NAI; "Minutes of the meeting held in the Office of The Braithwaite Burn & Jessop Construction Co. Ltd.... on Friday, the 29th January, 1960, at 11.00 A.M.," ibid.; Contract, "Manufacture, supply, erection and maintenance for one year of ten spans of 397′ feet [*sic*] and two spans of 104′ -9″ girders for the Brahmaputra Bridge at Amingaon," January 25, 1965, file no. 65/W5/BR1/2/1-9, Ministry of Railways, NAI; Ponnuswamy, "Challenges in Bridging River Brahmaputra," 400; Ponnuswamy and Sharma, *Bridging River Brahmaputra*, 21.

piers started at Amingaon on the north bank of the Brahmaputra and proceeded sequentially, one by one, southward across the river. This was better from a logistical standpoint because the north bank had the rail link with the Indian mainland, and a rock quarry was located near the project site in Amingaon. The bridge was designed to have eleven main piers founded in the riverbed and one shore pier each at Amingaon and Pandu. During the 1958–1959 working season, HCC completed the foundation wells of Piers 1 and 2 and the Amingaon side shore pier. It was critical that the wells were firmly founded in the riverbed before the pre-monsoon floods began, as a partially completed well could be knocked out of alignment, thus ruining the entire bridge.[43]

Using steel caissons supplied by Braithwaite, Burn and Jessop, HCC excavated deep foundation wells for the river piers. The foundations were built to be one foot higher than the highest flood level of the river. On top of each foundation well, HCC built twin concrete cylinders of sixteen-foot diameter that raised the level of the bridge an additional forty feet above the high flood level. HCC returned to the project in 1959–1960 and sank the wells of Piers 3 through 7, as well as the shore pier on the Pandu side. In the last season, 1960–1961, HCC completed the wells of the four remaining main piers, numbers 8 through 11.[44]

While sinking Piers 1 and 2 in the first season, HCC discovered to its dismay that the Brahmaputra riverbed had a much higher clay content than expected. The original 1958 railway-contractor agreement had been based on the experience of the Ganga Bridge at Mokameh Ghat. In his January 2 letter to the Railway Board, S.S. Godbole of HCC stated that the company understood that "wells will be sunk through strata consisting mostly of sand and very little of clay." The Ganga riverbed in Bihar had consisted of 2 percent clay. By contrast, the Brahmaputra riverbed was

[43] Memo, "Brahmaputra Bridge Contract," NFR to Rail Secretary, March 9, 1965, file no. 62/W5/BR1/17/1-135, Ministry of Railways, NAI; Karnail Singh, "Improvements in Rail Communications Between Assam and the Rest of India," memo, September 5, 1957, file no. 57/W5/BRI/17 (Contract)/1-135, Ministry of Railways, NAI.

[44] "Discussion Note on Technical Data for the Brahmaputra Bridge at Amingaon," November 1957, file no. 57/W5/BRI/17 (Contract)/1-135, Ministry of Railways, NAI; "Brief Technical Note on the Foundations of the Brahmaputra Bridge in Comparison with Those of the Ganga Bridge at Mokameh," n.d. [February 1958], ibid.; "Brahmaputra Conquered After Fifty Years' Planning," *Assam Tribune Magazine*, June 17, 1962; Ponnuswamy and Sharma, *Bridging River Brahmaputra*, 49–53. The chapters about Saraighat Bridge in *Bridging River Brahmaputra* are based on the papers of B.C. Ganguli.

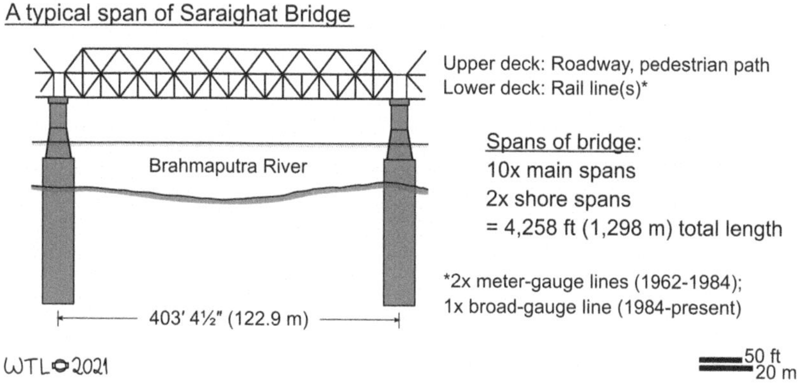

Fig. 6.3 Diagram of a typical span of the Saraighat Bridge (Drawing by the author, based on diagrams and data in S. Ponnuswamy and V.C. Sharma, eds., *Bridging River Brahmaputra: Past Present and Future* [Guwahati: Northeast Frontier Railway, 1988], esp. p. 30)

fully 29 percent clay. The higher clay content forced HCC to incur higher expenses, as it had to use heavy kentledge blocks[45] to weigh down the wells, and even resort to blasting in the case of some piers. HCC claimed that NFR owed it Rs. 180,000 on account of this additional work. NFR refused at first, but at length agreed to pay Rs. 170,000 after arbitration by the Railway Board.[46]

The wells and piers of the Brahmaputra Bridge were similar in many respects to those of the Ganga Bridge, but HCC had to reinforce the design to withstand the earthquakes that regularly rock Assam. The Brahmaputra Valley lies at a major geological fault between the Australian-Indian and Eurasian plates, and earthquakes occur frequently along this fault. The Ganga Bridge had been designed to withstand lateral acceleration of g/20, or 5 percent of acceleration due to the force of gravity

[45] Kentledge blocks are large rectangular blocks of a heavy material (usually concrete), used on construction projects for ballast or footings for temporary structures.

[46] S.S. Godbole to Rail Secretary, January 2, 1958, file no. 62/W5/BR1/17/1-135, Ministry of Railways, NAI; N.J. Vacha to NFR Works, December 1, 1960, ibid.; "Note on the Discussion Held on 16.7.1965 with M/s Hindustan Construction Co. Regarding Their Outstanding Claims in Respect of Brahmaputra Bridge Works," memo, n.d. [1965], ibid.

Fig. 6.4 The Sankosh River Bridge on the Assam Rail Link in 2015. The spans are single Warren trusses (Photo by the author)

near the Earth's surface. By comparison, NFR's requirements for both the piers and the trusses called for them to withstand a seismic factor of g/10, twice the design standard of the Ganga Bridge. HCC brought the design of the piers up to code by making the foundation walls thicker and using a richer mix of concrete. The mix of concrete had to be adjusted daily, depending on the moisture content of the aggregate, as tested in an on-site laboratory.[47]

Even before all the piers were completed, BBJ began erecting the trusses. Like the piers, the trusses were also similar in design to those of the Ganga Bridge, although they too had to be strengthened to withstand higher seismic loads. Additionally, they were redesigned to carry two rail lines on the lower deck. On both the Ganga and the Brahmaputra bridges, the spans were double Warren trusses (Fig. 6.3). Made from equilateral triangles placed side-by-side, Warren trusses are easy to construct because all girders are the same length and thus can be ordered in bulk (Fig. 6.4). A double Warren truss is simply one Warren truss

[47] NFR, Brahmaputra Bridge Project Report (1959); "Brief Technical Note on the Foundations of the Brahmaputra Bridge in Comparison with Those of the Ganga Bridge at Mokameh," n.d. [February 1958], file no. 57/W5/BRI/17 (Contract)/1-135, Ministry of Railways, NAI; Ponnuswamy and Sharma, *Bridging River Brahmaputra*, 51.

stacked on top of another. On these bridges, the upper truss carried the roadway, while the lower truss carried the dual meter-gauge rail lines.[48]

As with the piers, the erection of the trusses began on the north bank and proceeded southward. With the help of derrick cranes that traveled along the uppermost girders of the bridge, work crews built the trusses in place, piece-by-piece, cantilevering each new truss out over the water. As the incomplete trusses extended from one pier to the next, stresses developed in the girders that would not occur once the trusses were complete. Parts of the trusses had to be temporarily strengthened with extra material, which was removed when the trusses were complete.[49]

As construction progressed, ceremonies marked key stages of the project. On November 28, 1960, Karnail Singh of the Railway Board presided over the grounding of the caisson for the bridge's eighth pier. The chairman of the Thai Railway Board, who was touring the project with other engineers from southeast Asia, also participated in the inauguration. The representatives of the Indian and Thai railway boards each broke open a coconut as part of the inauguration. A coconut also featured prominently in the inauguration of the bridge's final truss on July 14, 1960. After B.C. Ganguli broke a coconut, a crane lifted the last girder section into place. In the words of the *Assam Tribune's* report on the event, on Ganguli's signal "the aluminum erection crane hoisted it gracefully to the jubilation of all those present." These milestone ceremonies drew on a Hindu tradition of breaking coconuts at the beginning and end of construction projects, thereby linking this modern project, built along foreign lines, with India's cultural and religious heritage.[50]

On October 31, 1962, as Indian and Chinese troops were fighting each other in the mountains to the north, B.C. Ganguli presided over a ribbon-cutting ceremony that declared the bridge officially open to goods traffic. At this point, even though one rail line was open on the bridge, much work remained to be done. The second rail line on the lower deck and the roadway on the upper deck were not completed yet. Passenger trains started crossing the bridge in December, and motor vehicles began to use the roadway in March of the following year. Later, an oil pipeline

[48] "Ponnuswamy, Challenges in Bridging River Brahmaputra," 400–1; Marshall M. Kirkman, *Building and Repairing Railways* (New York: World Railway, 1907), 271.

[49] Ponnuswamy and Sharma, *Bridging River Brahmaputra*, 53.

[50] "ECAFE Engineers at Pandu: Visit to Brahmaputra Bridge Sites," *AT*, November 29, 1960; "Brahmaputra Bridge: Erection of Girders Completed," *AT*, July 15, 1962.

would also be installed on the bridge, connecting the oil fields of upper Assam with the Indian mainland.[51]

On a rainy June day seven months after the bridge opened, Prime Minister Nehru visited Assam to dedicate the bridge. At the ceremony held at Maligaon Railway Stadium, Nehru announced to the crowd of 10,000 that the name of the bridge would be Saraighat, after the battle in which the Ahoms defeated the Mughal army in 1671. According to the *Assam Tribune*, "The Prime Minister's announcement was greeted with loud cheers and hand-clapping by the crowd who collected in their thousands at the Maligaon Railway Stadium despite continuous rain."[52]

Despite the name, Saraighat Bridge was ultimately more Indian than it was Assamese. The project was financed by the central government, directed by a Bengali engineer, overseen by the Railway Board represented by a Punjabi engineer, and constructed by firms based in Calcutta and Bombay using steel from Bihar and the labor of non-Assamese Indian workers.[53] Even the name of the bridge had been chosen not by an Assamese, but by an outsider, the North Indian Prime Minister. Although the bridge was named for an event that had ensured, for a time, Assam's independence from the pan-Indian Mughal Empire, independence for Assam was now out of the question. Rather, Nehru's selection of the name in the modern period reflected his belief in plurality and regional distinctiveness within the united Indian nation (Fig. 6.5).

[51] "Saraighat Bridge: A Boon to Assam Stands to Serve the Nation," *Assam Information* 1963, 19–20; J.N. Roy to B.D. Gaur, November 4, 1960, file no. 57/W5/BRI/17 (Contract)/1-135, Ministry of Railways, NAI; M.M. Kohli to D.R. Kohli, November 2, 1960, ibid.; "Rail Bridges to Carry Pipelines," *Hindustan Times*, June 29, 1963. Also crossing the bridge are various cables and battered old telegraph lines.

[52] "Brahmaputra Bridge Named Saraighat," *AT*, June 8, 1963; "Brahmaputra Bridge Inaugurated," *India News*, June 14, 1963.

[53] "Consideration of the report of the Commissioner for Scheduled Castes and Scheduled Tribes," *Assam Legislative Assembly Debates* 1 (1959): 3433–34; E.H. Pakyntein, *Kamrup District Census Handbook, Census of India 1961* (Government of Assam, 1964), Assam State Archives.

Fig. 6.5 Saraighat Bridge in 2015 (Photo by the author)

Conclusion

When it opened in 1962, Saraighat Bridge was the first bridge across the Brahmaputra River, but it was not the last. The Kalia Bhomara Bridge in Tezpur, which opened in 1987, and the Narayanan Setu at Jogighopa, completed in 1998, cross the river at other strategic locations. The Jogighopa Bridge, 65 miles downstream from Saraighat Bridge, provides improved connectivity for lower Assam. If you had felt so inclined on your imaginary rail journey to Assam, you could have gotten off the train at New Bongaigaon station and taken a bus across the Jogighopa bridge to Goalpara or Krishnai, and then caught a van or a sumo up into the Garo Hills.[54]

In 2002, the Northeast Frontier Railway started building the fourth bridge over the Brahmaputra, at a site near Dibrugarh in upper Assam. Bogibeel, as the bridge is called, is designed to increase connectivity in Assam and provide a strategic link with the border areas of Arunachal Pradesh. Gammon India received the contract to build the bridge's foundations and piers, while the contract for the welded steel trusses was awarded to a consortium of three companies: Hindustan Construction Co., VNR Infrastructures of Hyderabad, and a German firm, DSD Brückenbau GmbH. The German firm was responsible for the design of the trusses.[55]

[54] Ponnuswamy, "Challenges in Bridging River Brahmaputra," 401–5.

[55] "HCC, Associates Bag Rs 987-cr Railway Order," *The Hindu BussinessLine*, November 24, 2011; "Work on Bogibeel Bridge Progressing Well, Claims NFR," *AT*, July 6, 2014.

Note the difference between Saraighat and Bogibeel Bridges. In 1959, Saraighat Bridge's trusses were designed by an Indian company, with only a little help from a British subcontractor. In 2011, the trusses for the next-generation Brahmaputra Bridge were designed in Germany by a German company. In 1959, bridge construction practices in India were up-to-date with those abroad, but this was no longer the case in 2011.

The trusses of the Bogibeel Bridge demonstrate that technology transfer into a country is a continual process, not a single event. One-time transfer can lead to a technology being seemingly frozen in time in one country as the technology continues to change abroad. For an example of this approach, look no further than the Hindustan Ambassador, the preferred ride of *netas* (Indian politicians). Manufactured near Calcutta by Hindustan Motors (HM), the Ambassador was based on the Morris Series II Oxford, which was manufactured in England starting in 1954. Using secondhand tooling from Morris, HM set up the production line of the Ambassador later in the fifties. The Ambassador got a new engine in 1959, but otherwise the cars were produced nearly unchanged for more than twenty years. (Government regulations prohibited technology transfer for established automobile production lines.) During this time, automotive technology—and not to mention automotive styling—changed considerably outside of India.[56]

This was one approach to technology transfer and indigenization. Another approach was piecemeal importation of technologies, as had been done for the Brahmaputra bridge projects. A third approach would be for India to become a major technological innovator in its own right, but this has so far failed to happen in most fields. One-time technology transfer could hardly be satisfying to a nation with great industrial ambitions. For these reasons, technology transfer has been a continual process. Technology transfer was hardly necessary for the first Brahmaputra Bridge, but by the beginning of the twenty-first century, technology outside India had changed enough to make importation of foreign technology desirable.

[56] *The Complete Encyclopedia of Motorcars: 1885 to the Present* (2nd ed., New York: E.P. Dutton, 1973), 358; Ziauddin Sardar, "The Ambassador from India," in *Autopia: Cars and Culture*, eds. Peter Wollen and Joe Kerr (London: Reaktion Books, 2002), 208–18; R.C. Bhargava and Seetha, *The Maruti Story: How a Public Sector Company Put India on Wheels* (Noida, India: Collins Business, 2010), 4. Another Indian car that was also frozen in time was the Premier Padmini, based on a Fiat design and produced in Bombay. Padmini taxis are still a common sight on the streets of Mumbai.

Let us return to the image that opened this chapter: your imaginary rail journey to Assam. The journey represents the great changes that have taken place in Assam since independence and Partition. Traveling by train is one way to observe these changes, but there are many others. You could travel by car or bus; or take a boat on the Brahmaputra; or simply walk down the streets of Guwahati, a provincial town like Goalpara, or a village. Everywhere you would see evidence of postindependence modernization and development: roads, rail lines, buildings, and power lines.

In the final three chapters of this book, we will consider one of these aspects of modernization: electric power. Perhaps no other technology is as emblematic of twentieth-century modernity—or as representative of its empty promises—as electric power. For the next chapter, we will not have to go far from the Brahmaputra Bridge, as we will be traveling up into the Khasi Hills southeast of Guwahati to study the Umiam Hydroelectric Project.

But let us linger by the banks of the Brahmaputra for a moment before heading up into the hills. After the Saraighat Bridge opened, the Amingaon–Pandu ferry shut down, and the locomotive sheds on both sides of the river were shifted to the new rail yard at Narangi. Automobile ferries also shut down, as cars were rerouted onto the road deck of the new bridge. As Gauhati and the neighboring towns grew, traffic increased across the bridge. Today the bridge hosts frequent traffic jams during rush hours.[57]

Even though the bridge eliminated the need for train and automobile ferries, it did not put all ferries in the Guwahati area out of business. Located as it is five miles downstream from Guwahati, Saraighat Bridge is awkwardly sited to serve the city. The quickest route across the river from Guwahati to North Guwahati (home to a campus of the Indian Institute of Technology) is still by ferry. Small wooden ferries, operated by the state government's Inland Water Transport Department, still make regular crossings of the broad expanse of the river. The boats are big enough to carry scooters or motorbikes, but not automobiles. On either side of the river, the ferry docks are themselves decommissioned ferries, made of metal and much larger than the craft currently operating. Some have landing ramps and a deck large enough to hold several cars. They are relics of a time when all traffic crossed the river by boat.

[57] N. Krishnaswamy to Y.P. Kulkarni, January 28, 1960, file no. 59/W5/BR1/19/1-22, Ministry of Railways, NAI.

References

1. Bhargava, R.C., and Seetha. *The Maruti Story: How a Public Sector Company Put India on Wheels*. Noida, India: Collins Business, 2010.
2. "Bridge Site: Jogighopa, Goalpara District, Assam." *Records of the Geological Survey of India* 79 (1954): 820.
3. Dutta, Arup Kumar. *Indian Railways: The Final Frontier—Genesis and Growth of the North-East Frontier Railway*. Guwahati: Northeast Frontier Railway, 2002.
4. Gait, Edward. *A History of Assam*. 1906. Reprint Delhi: Surjeet Publications, 2006.
5. Goswami, Priyam. *The History of Assam from Yandabo to Partition, 1826–1947*. Hyderabad: Orient Blackswan, 2012.
6. Headrick, Daniel R. *The Tentacles of Progress: Technology Transfer in the Age of Imperialism, 1850–1940*. New York: Oxford University Press, 1988.
7. Kailasam, L.N. "Some Results of the Application of Resistivity Methods to Problems of Civil Engineering, Mining and Ground Water in India." *Bulletins of the Geological Survey of India*, Series B: Engineering Geology and Ground-Water, no. 3. Calcutta: Government of India Press, 1952.
8. Kirkman, Marshall M. *Building and Repairing Railways*. New York: World Railway, 1907.
9. Medhi, Shyam Bhadra. *Transport System and Economic Development in Assam*. Gauhati: Assam Publication Board, 1978.
10. Neog, Maheswar. *Lachit Barphukan: The Victor of the Battle of Saraighat*. New Delhi: Publications Division [Government of India], 1983.
11. Piramal, Gita. *Business Legends*. New Delhi: Penguin Books, 1998.
12. Planning Commission. *First Five Year Plan*. http://planningcommission.nic.in/plans/planrel/fiveyr/1st/welcome.html (accessed January 13, 2014).
13. Planning Commission. *Second Five Year Plan*. http://planningcommission.nic.in/plans/planrel/fiveyr/2nd/welcome.html (accessed January 13, 2014).
14. Ponnuswamy, S. "Challenges in Bridging River Brahmaputra." *Indian Concrete Journal* 71, no. 8 (August 1997): 399–405.
15. Ponnuswamy, S., and V.C. Sharma, eds. *Bridging River Brahmaputra: Past Present and Future*. Guwahati: Northeast Frontier Railway, 1988.
16. Saikia, Arupjyoti. "Ecology, Floods and the Political Economy of Hydropower: The River Brahmaputra in the 20th Century." In *Perspectives in Indian Development: New Series* (New Delhi: Nehru Memorial Museum and Library, 2013), 2:1–46.
17. Saikia, Arupjyoti. "Jute in the Brahmaputra Valley: The Making of Flood Control in Twentieth-Century Assam." *Modern Asian Studies* 49, no. 5 (September 2015): 1405–41.

18. Sardar, Ziauddin. "The Ambassador from India." In *Autopia: Cars and Culture*, edited by Peter Wollen and Joe Kerr, 208–18. London: Reaktion Books, 2002.
19. Singh, Jagman. *My Tryst with the Projects Bhakra and Beas*. New Delhi: Uppal Publishing House, 1998.

CHAPTER 7

A Dam in the Khasi Hills: The Umiam Hydroelectric Project, 1960–1965

In early 1960, the Assam State Electricity Board placed an order for four 560-kW diesel generators from the Czechoslovakian firm STC. The state electricity board intended to use some of the generators at the construction site of the Umiam (or Barapani) Hydroelectric Project, which it had just started building in the Khasi Hills of Assam. By itself this order was totally unremarkable. Indian development projects relied heavily on eastern-bloc equipment, which was often cheaper and more easily accessible than equipment from the West. Entire factories for import substitution were being set up in India with Czechoslovakian machinery. In Gauhati, down the hills from the dam site, the Noonamati refinery was being built with Romanian assistance.[1]

In context, though, this order was quite significant. The Umiam project was being financed by the US government under the auspices of the Development Loan Fund. According to DLF policy, American-financed projects could not use equipment from countries designated as foreign policy-restricted under Code 899 of the ICA Geographic Code

[1] E. Kolet, Extract of CWPC note recorded in file no. F.2/4/50-DW.III, May 3, 1960, file no. 15(11)60, Ministry of Finance, NAI; C.S. Swaminathan to S.S. Kumar, June 21, 1960, ibid.; "Starting Production of Heavy Power Equipment in India," *Czechoslovak Heavy Industry*, May 1966, 20; "Nehru Inaugurates Gauhati Refinery This Morning," *Assam Tribune*, January 1, 1962.

W. A. T. Logan, *A Technological History of Cold-War India, 1947–1969*, Palgrave Studies in the History of Science and Technology,
https://doi.org/10.1007/978-3-030-78767-7_7

Book. As explained in DLF Implementation Letter No. 2, countries not on the American-approved list were the "Union of Soviet Socialist Republics, Albania, Bulgaria, Czechoslovakia, Soviet-occupied Germany, Hungary, Poland, Rumania and China (Mainland)."[2]

The engineers at the state electricity board did not think of the broader geopolitical ramifications when they placed a seemingly innocuous order for diesel generators. But the bureaucrats at the central government Finance Ministry did think in these terms when they learned about the order. After hearing about the order, C.S. Swaminathan of the Finance Ministry inquired of the Central Water and Power Commission whether the diesel sets were integral to the project. K.L. Vij of CWPC replied that in his opinion, the diesel sets were not integral, as they would be removed immediately after completion of construction. Swaminathan nevertheless insisted that the state electricity board not use any eastern-bloc equipment on this project. As he explained in a memo, "It appears that we should not take a risk of allowing equipment from countries not acceptable to DLF to be used even in the construction stage. By doing so (even if the action may be deemed technically defensible), we will not be strengthening the confidence of DLF in us." The state electricity board agreed to put the Czechoslovakian sets to use elsewhere, and instead placed new orders in Yugoslavia, which was a nonaligned nation and therefore not considered foreign policy-restricted by the Americans.[3]

The story of the Assam State Electricity Board and the Czechoslovakian generators shows how development projects—even smaller-scale ones of only regional significance like Umiam—were built in the context of the Cold War. Development was itself a cold-war weapon deployed by both sides, but especially the United States. The Cold War restricted access to technology for development projects, just as it did for the Indian Air Force and its fighter acquisitions (see Chapter 4). Sometimes the restriction was minor and only a few bureaucrats and engineers in India were even aware of it. At other times, incidents far removed from the

[2] C.S. Swaminathan, handwritten note on memo by T.V. Thadani about DLF assistance to Barapani project, June 15, 1960, file no. 15(11)60, Ministry of Finance, NAI; E. Kolet, Extract of CWPC note recorded in file no. F.2/4/50-DW.III, May 3, 1960, ibid.

[3] C.S. Swaminathan, memo, July 2, 1960, file no. 15(11)60, Ministry of Finance, NAI; K.L. Vij to C.S. Swaminathan, June 24, 1960, ibid.; C.S. Swaminathan to K.L. Vij, July 8, 1960, ibid.

project could lead to a very public denial of aid that made headlines around the world.[4]

The Umiam Hydroelectric Project was financed by the United States almost from the very beginning—a first for an Indian hydroelectric project. A few earlier projects had already been underway when they received American financing, but the Umiam Project was just starting when the US government offered two loans to finance construction from start to finish. The first was a dollar loan of $2.5 million, meant to cover foreign-exchange costs incurred while importing goods or paying foreign experts. The second loan, falling under Public Law 480 and worth Rs. 64 million ($13.1 million at the time), covered domestic costs.[5]

In addition to financing, tools and specialized components for building Umiam Dam also had to be imported. American industry provided earth-moving machinery and instrumentation. Japan provided four turbine generators, as well as financing for the fourth unit, a later addition that was not covered by the American loan. Other components came from Austria, France, West Germany, Italy, Sweden, and the United Kingdom. There were also, of course, the diesel generators from Yugoslavia.[6]

Even though the Umiam Hydroelectric Project was American-financed and built on design principles that originated in the West, the project was Indian in conception, design, and implementation. The Assam State Electricity Board (ASEB) built the project departmentally. Khasi landowners

[4] The highest-profile example of denied American aid to India was Stage 1 of the Bokaro Steel Plant. At $512 million of foreign exchange, this was the largest single aid package that India had yet requested from the United States. President Kennedy was in favor of American financing of the plant, and Ambassador Galbraith all but promised Nehru's government that it would receive the requested financing. But the Republican-controlled House of Representatives blocked the aid appropriation bill on grounds that the American government should not support a project that might compete with US private industry. The plant was ultimately built with Soviet aid. US House, Committee on Appropriations, Foreign Operations Appropriations for 1964, Hearing, July 22, 1963 (Washington: Government Printing Office, 1963), 1396; ibid., Hearing, July 24, 1963, 1481; Padma Desai, *The Bokaro Steel Plant: A Study of Soviet Economic Assistance* (New York: American Elsevier Publishing Company, 1972), 39–41.

[5] DLF, press release on loan for Barapani project, July 8, 1960, box 2, USAID Mission to India Subject Files 1955–1962, RG 286, NARA; Allen J. Ellender, *Review of United States Government Operations in South Asia* (Washington, DC: Government Printing Office, 1968).

[6] T.V. Thadani to C.L. Garg, August 28, 1962, file no. 15(11)60, Ministry of Finance, NAI.

sold the land taken up by the dams, powerhouse, and reservoir. The Planning Commission placed the project in the Second and Third Five-Year Plans. The Central Water and Power Commission (CWPC) prepared the plans at its headquarters in Delhi and tested designs of the dam in its laboratory in Poona. Domestic contractors provided construction materials and even a few of the specialized electrical components.

By 1960, the engineers of the Central Water and Power Commission and the Assam State Electricity Board already knew how to design and build dams. The CWPC had long institutional roots in the colonial era, and many of its and the ASEB's engineers had degrees from foreign universities. Only certain specialized knowledge, like the use of instrumentation to monitor the dam, had to be imported for this project.[7]

This chapter uses the Umiam project as a case-study for understanding dam-building in Nehruvian India. For Umiam, as for many comparable dams built in the early independence period, expertise was not the limiting factor. Even more than for the HF-24 Marut, the limiting factor was economic. The Indian workforce had the expertise to design a hydroelectric project, but the Indian economy could not provide the capital to finance this particular project, nor could it manufacture the machinery needed to build it.

Dams in Independent India

The early independence period was an era of prolific dam-building in India. Umiam Dam was one of 648 dams higher than ten meters (33 feet) built in India between independence and the beginning of the Fourth Five-Year Plan in 1969. These dams were built to harness nature and make it serve the needs and wants of humanity—or some subset thereof—by storing or diverting water for irrigation, generating electricity, or supplying water to India's thirsty cities.[8]

[7] Satyajit Singh, *Taming the Waters: The Political Economy of Large Dams in India* (Delhi: Oxford University Press, 199), 52; Daniel Klingensmith, *"One Valley and a Thousand": Dams, Nationalism, and Development* (New Delhi: Oxford University Press, 2007), 70.

[8] Central Water Commission, "National Register of Large Dams—2009," http://www.cwc.nic.in/main/downloads/National%20Register%20of%20Large%20Dams%202009.pdf (accessed October 25, 2012).

India has a very long history of hydraulic engineering, long predating the arrival of the British and their technology. In premodern Assam, the Ahom kingdom used a forced-labor system to build reservoirs, embankments, and temple tanks. To the west of the Khasi Hills, the Garos of the western Meghalaya Plateau built weirs and other structures to trap fish in their rivers. Throughout India, temporary diversions were set up in rivers to direct water into irrigation channels at the appropriate time. There were also many large reservoirs built with permanent dams, especially in arid central and western India. The largest extant premodern artificial lake in India is Jaisamand, thirty miles south of Udaipur in southern Rajasthan. Built in 1685 on behalf of Maharana Jai Singh of Mewar, the twenty-square-mile lake is impounded by a 1,500-foot marble dam.[9]

There was not much continuity between the hydraulic engineering of premodern India and the dams and canals built in the colonial period and after independence. The British brought reinforced-concrete construction techniques and heavy earth-moving machinery, which allowed dams to be built more quickly and cheaply than by traditional methods. They also introduced hydroelectricity. India's first small hydroelectric plant, located near the hill station Darjeeling, came online in 1897. Five years later, the state of Mysore built a much larger project at Sivasamudram Falls. After World War II, electrification expanded in India, in conjunction with industrialization. Power stations were built all across India, including a small hydroelectric plant at Bishop's Falls near Shillong. Power supply remained localized, focused on urban centers with limited interconnection between grids.[10]

Electrification had slowed during World War II, as industry retooled for military production, but then it picked up again after independence as the five-year plans set high goals for industrialization. Dams, of course,

[9] Singh, *Taming the Waters*, 25, 34; James Tod, *Annals and Antiquities of Rajasthan of the Central and Western States of India* (London: Oxford University Press, 1920), 1:458; *Large Dams in India* (New Delhi: Central Board of Irrigation and Power, 1987), 1:206–7; "Vast Marble Dams, Hidden for Centuries in Hills of India, Are Unsung Triumphs of Engineering," *Popular Science Monthly*, April 1922, 37–38.

[10] V. Venkatarama, *Power Development in India: The Financial Aspects* (New Delhi: Wiley Eastern, 1972), 29; United States Department of Commerce, *The Electrical Equipment Market in India* (Washington, DC: Government Printing Office, 1927), 8; Hamlet Bareh Ngap Kynta, *The Economy of Meghalaya: Tradition to Transition* (Guwahati: Spectrum Publications, 2001). The original Swiss-made Pelton turbine runner from the powerhouse of the Sheshadri Iyer Power Station at Sivasamudram Falls is on display in the Visveswaraya Industrial and Technological Museum in Bangalore.

were a significant part of postindependence electrification and development. Between 1947 and 1969, 52 dams were built in India for power generation. One of these dams impounded the cooling water reservoir for Satpura Thermal Power Station, while the remaining 51 belonged to projects with hydroelectric powerhouses. Although these 52 dams were greatly outnumbered by dams made primarily for irrigation, they tended to be bigger in actual size as well as public profile. Hydroelectric dams need to be tall to produce a head for powering turbines or impounding a reservoir. (Twelve of the dams with powerhouses were multipurpose and also served as headworks of irrigation systems.)[11]

Agencies of the Indian central and state governments initiated the majority of dam projects in the early independence period. One agency that got a considerable amount of attention was the Damodar Valley Corporation, which was the joint responsibility of the central government and the state governments of West Bengal and Bihar (later Jharkand). Envisioned as a clone of the Tennessee Valley Authority of the United States, DVC quickly shifted from building dams to thermal power stations. Elsewhere, state electricity boards and water resources departments built dams for hydroelectricity, irrigation, or mixed use. City governments also built dams, for storing municipal water supplies. Dams could be constructed departmentally, as in the case of Umiam Dam, or by a private contractor like Hindustan Construction Co.[12]

In addition to providing contractors and materials, Indian private industry also advocated the continued development of rivers in India. A mouthpiece of the private-sector hydroelectricity lobby was the *Indian Journal of Power and River Valley Development*, which commenced publication in Calcutta in 1951, the same year the First Five-Year Plan began. The journal carried technical articles reporting the research of Indian engineers and scientists, as well as more generally accessible articles about power and irrigation development in India and abroad. The journal also brought out frequent special issues to celebrate the completion of

[11] Singh, *Taming the Waters*, 94. Data from Central Water Commission, "National Register of Large Dams—2009," http://www.cwc.nic.in/main/downloads/National%20Register%20of%20Large%20Dams%202009.pdf (accessed October 25, 2012).

[12] Hindustan Construction Co., "Directors' Report and Statement of Accounts for the Year Ended 31st August 1950," S. No. 340, Hindustan Construction Company subject files, papers of Walchand Hirachand, NMML; *Large Dams in India*, 1:205–13; Klingensmith, *One Valley and a Thousand*, 110; "Overview," *Damodar Valley Corporation*, http://www.dvc.gov.in/dvcwebsite_new1/overview/ (accessed November 24, 2015).

major dams or to cover broader topics such as electrical development in particular states.[13]

The development of hydroelectric projects in the early independence period was concentrated in southern India. Of the fifty-two dams built for power generation between 1947 and 1969, more than half of them were located in the two southernmost states of the country, Kerala and Madras (Tamil Nadu). The second-largest hydroelectric project from this period—second only to the headline-hogging Bhakra-Nangal in Punjab—was the Sharavathi Hydroelectric Project, built in the southern state of Karnataka with partial US government financing.[14]

Two hydroelectric plants were built in Assam in the early independence period, and both were small compared with the giant, high-profile projects built in other parts of the country. The first of these was the Umtru Hydroelectric Project, built by the Assam Electricity Department in the lower flanks of the Khasi Hills. The project consisted of a weir that diverted water from the Umtru River into a 4,600-foot conduit running to a powerhouse. Inside the powerhouse were three turbine generators of 2.8 MW each. The Central Water and Power Commission of the Indian government prepared plans for the Umtru project, and Montreal Engineering Company served as consultant. The project also received financial aid from the Dominion of Canada under the Colombo Plan.[15]

The second project, of course, was Umiam Hydroelectric Project. The first explosive charges of the project were set off by Prime Minister Nehru on January 9, 1960, and the US government announced six months later that it would finance the project in its entirety.

A Dam on Khasi Land

American financing would be essential for building Umiam Dam, and it may be that the project would never have been completed without it. But there was much more besides financing that the project needed

[13] Klingensmith, *One Valley and a Thousand*, 34; Singh, *Taming the Waters*, 52.

[14] "National Register of Large Dams—2009"; H. Kusumakar, "Country's Largest Single Powerplant," *TOI*, January 24, 1965.

[15] "Barapani Hydro-Electric Project (Assam): Installation of IV Generating Unit," n.d. [August 1961], file no. 6(16)/61 vol. 3, Ministry of Finance, NAI; "India-Canada Co-operation in River Valley Development," *Indian Journal of Power and River Valley Development* (*IJPRVD*) 12, no. 10 (October 1962): 19–20, 26.

before it could be completed. It needed materials, equipment, and labor. And it needed land. In the 1950s, before construction on Umiam Dam began, the land that would be flooded by Umiam Lake was occupied and owned by people of the Khasi tribe. The Khasis are registered as a Scheduled Tribe under the Sixth Schedule of the Indian Constitution, which means that they are an ethnic group that falls outside the Indo-Aryan or Dravidian mainstream of Indian society. The Khasis' Mon-Khmer language is unrelated to either Sanskrit or the Dravidian languages of South India. Although no historical records exist of their arrival in India, their language suggests that they originated in Southeast Asia. The Khasis were a preliterate society before the nineteenth century, but they have a wide range of oral traditions that describe their past.

Most subdivisions of the Khasi tribe are both matrilineal and matrimonial, which means that property and the family name pass through the mother to her children. The name Khasi means "born of the mother," and mothers traditionally control the household and manage family business. But Khasi society is not matriarchal, as men dominate administration. The leaders of Khasi society are the (mostly male) chiefs known as *syiems* and *sardars*. They preside over darbars (village councils), which are composed of all male residents of an area. The *syiems* and *sardars* control the disposition of public lands, although traditionally they have no authority to make final decisions on land transfer without approval of the darbar. The Khasis originally recognized three types of land-ownership: *Ri-Raid* (public land), *Ri-Kynti* (land belonging to a single clan), and sacred groves, which were set aside for ceremonial purposes only.[16]

The arrival of British colonists in the first half of the nineteenth century added another category to the Khasi land tenure system: government land. The British acquired some land inside the Khasi syiemships for their own use, such as at the hill stations of Cherrapunjee and Shillong. Unlike in the neighboring Jaintia Hills, the Khasi states were not annexed outright by the British. Rather, the colonialists recognized the traditional territorial rights exercised by the *sardars* and *syiems*, and accordingly treated the Khasi *syiemships* as autonomous princely states. Toward the

[16] Hamlet Bareh, *The History and Culture of the Khasi People* (published by the author; printed by Naba Mudran, Calcutta, 1967), 250, 287, 337; J.N. Das, *A Study of the Land System of Meghalaya* (Guwahati: Law Research Institute of Gauhati High Court, 1990), 54–56.

end of the nineteenth century, the British recognized twenty-five different states within the Khasi Hills.[17]

In the years immediately after independence and Partition, the Khasis transitioned to the status of minorities with protected rights within the Indian state. Occupying territory on the new border between India and East Pakistan, the Khasi states had the choice to accede to either country, and in fact the *syiems* of some southern states apparently did try to join Pakistan in order to maintain access to the traditional markets for their goods in eastern Bengal. But their letter got lost in the chaos of Partition, and by the time the Pakistani government got around to replying to it, the *syiems* had long since acceded to India instead. Thus it was that all of the Khasi Hills joined the Indian Union, and the Khasi people became a Scheduled Tribe after the constitution came into effect in 1950. Under the Sixth Schedule, the Khasis were empowered to set up Autonomous District Councils, which would manage the use of tribal lands. These councils were vested with the authority to sell land for development.[18]

In the mid-1950s, most of the land of the Umiam Valley was under tribal ownership. The single-lane Gauhati–Shillong Road, which could only carry traffic in one direction at a time, ran through the valley, and its right-of-way was owned by the government. (Timed gates at the top, middle, and bottom of the road allowed traffic to travel in only one direction on a stretch of road at a time.) Apart from the road, the valley was *Ri-Kynti* (public) and *Ri-Raid* (clan) tribal land.[19]

Shortly after its inception in June 1958, Assam State Electricity Board started to study in earnest the problem of damming the Umiam River for a hydroelectric project. The plan envisioned inundating part of the Umiam River Valley, where Khwan (Barapani) and Umsaw villages were

[17] Das, *A Study of the Land System of Meghalaya*, 53, 102, 104–6.

[18] Yaqoob Khan Bangash, *A Princely Affair: The Accession and Integration of the Princely States of Pakistan, 1947–1955* (Karachi: Oxford University Press, 2015), 143.

[19] C.S. Booth, timings on the Gauhati-Shillong Road, January 3, 1961, in *The Assam Gazette Part IIA*, January 11, 1961, 173; "Bus Accidents in Shillong-Gauhati Road," *Proceedings of the Assam Legislative Assembly*, May 2, 1958, in *The Assam Gazette Part VI*, January 20, 1960, 3649.

located. In total, about two hundred families would be forced to vacate their lands to make way for the reservoir, dam, and powerhouse.[20]

Not surprisingly, the residents of Khwan and Umsaw opposed the plan to flood their villages. In response, ASEB issued a public notice, printed in the *Assam Tribune*, justifying the project. ASEB listed the numerous benefits of the project—not just electricity but also irrigation, tourism, and pisciculture (fish farming), as well as creating jobs during construction. Despite the imploring prose of the notice, ASEB claimed that "in case the people do not want the Project, then the Project would not be taken up."[21]

Yet even though the people did not want the Project, the Project was taken up. On November 16, 1959, before the district council had given consent to the land acquisition, the Assam State Revenue Department issued a notice declaring that 2,400 acres of land in the Umsaw, Khwan, Mawlyndep, and Umniuh villages were "likely to be needed for a public purpose, viz., for the construction of Umniam [*sic*] Hydel Project by Assam State Electricity Board." Employees of the electricity board were authorized to enter the land and begin their surveys. A month later, on December 12, the tribal leaders, following the recommendation of an ad hoc committee, decided to acquiesce to the government and sell their land for the project.[22]

The negotiations between the government and the landowners were mediated by Stanley D.D. Nichols-Roy. Of mixed Khasi and American parentage, Nichols-Roy had been educated in Calcutta and the United

[20] M. Deka, notice of land acquisition in Shillong subdivision, November 16, 1959, in *The Assam Gazette*, November 18, 1959, 4089.

[21] "Controversy Over Barapani Project: State Electricity Board Clarifies Position," *AT*, October 27, 1958. Residents of other parts of Assam were unhappy that the power project was not being built in their district. In the Assam Legislative Assembly on September 15, 1959, Radha Krishan Khemka of Tinsukia (upper Assam) suggested that a hydroelectric project be built on the Dihing River: "The Dihing River also has the capacity to produce a very big measure of electricity, and if electricity were produced from that river, then Tinsukia, Makum, and other places nearby would have convenient access to electric power. At this time, Tinsukia and Makum need about 6,000 KW of power. Hopefully the government will make the arrangements to fulfill this requirement." "Discussion on the Budget of the State Electricity Board," *Assam Legislative Assembly Debates*, September 15, 1959, 237 (translated from Hindi by author).

[22] M. Deka, notice of land acquisition in Shillong subdivision, November 16, 1959, in *The Assam Gazette*, November 18, 1959, 4089; "Barapani Hydel Project: Khasis Now Give Consent," *AT*, December 17, 1958.

States before returning to the Khasi Hills to run fruit and road transport businesses. His father, the Reverend James Joy Mohon Nichols-Roy, had served in the government of Assam for decades and also made a name for himself by publishing a series of books about Jesus and Christianity. (Many Khasis had converted from their tribal religion to Christianity after coming into contact with Protestant missionaries in the nineteenth century.) As a highly educated, middle-class native of the Khasi Hills, the junior Nichols-Roy acted as a link between the worlds of the modern, non-tribal government, and the tribal landowners.[23]

The people displaced by the Umiam project were set to receive compensation for the loss of their land. Policy for rehabilitation of displaced persons had been developed in cooperation between the central and state governments. In 1957–1958, the Commissioner for Scheduled Castes and Scheduled Tribes had circulated recommendations among the state governments. One of the recommendations was that displaced persons should not be compensated in cash; instead, they should be "rehabilitated in well-planned colonies to be set up for this purpose." In its response to the recommendation, the Assam state government noted, "Government do not consider it desirable to segregate the Scheduled Caste and the Tribal people from the general communities lest it may defeat the aim of a classless society." This policy, while intended to improve the quality of life of the displaced people, nevertheless resembles coercive integration and assimilation programs undertaken by other governments against their tribal populations[24] (Fig. 7.1).

[23] Robert Gilkey, "Assam's 'Weeping River' Power Project," *AT*, June 18, 1961; Arthur J. Dommen, "Separatist Tendencies in Eastern India," *Asian Survey* 7, no. 10 (October 1967): 736; Frank Moraes, ed., *The Indian and Pakistan Year Book and Who's Who 1952–1953* (Bombay: Times of India, 1952), 760. Not long after brokering the Umiam project land sale, Nichols-Roy followed his father into politics by accepting a nomination to the position of General Secretary of the All Party Hill Leaders' Conference, the body that pressed for the secession of Meghalaya from Assam. After Meghalaya split from Assam in 1970, Nichols-Roy served as State Minister for Industries.

[24] "Statement showing action taken or proposed to be taken on the recommendations made by Commissioner regarding the rehabilitation of Scheduled Caste and Scheduled Tribe persons displaced on account of construction of industrial and agricultural projects in various states," in L.M. Shrikant, *Report of the Commissioner for Scheduled Castes and Scheduled Tribes for the Year 1959–60*, Part II, 166. Contrary to the central government's recommendation, the Assam state government did pay out cash compensations to the people displaced by the Umiam Project. In March 1961, Kamakhya Prasad Tripathi, Minister of Power of the state government, claimed that ASEB had already paid out almost

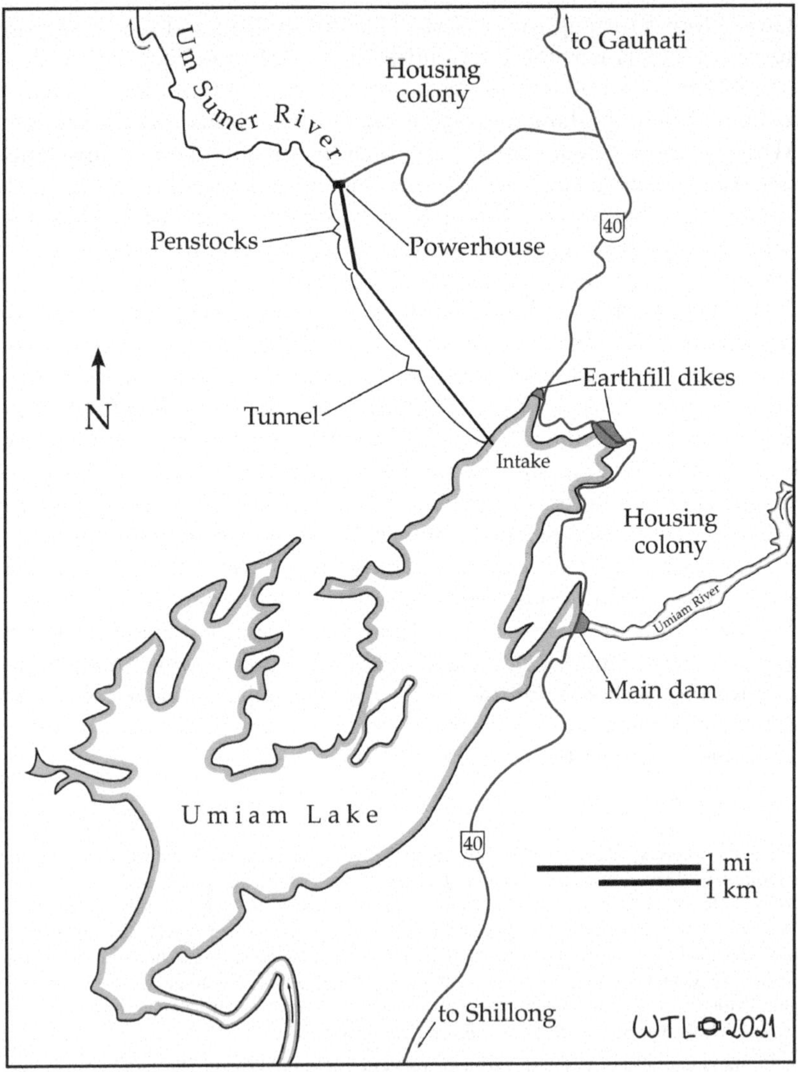

Fig. 7.1 Map of Umiam Lake and surroundings (Map drawn by the author, based on map in Assam State Electricity Board, "Umiam Hydro-Electric Project," commemorative pamphlet, n.d. [1966], Office of Director [Generation], Meghalaya Energy Corporation Ltd., Shillong)

How the Dam Was Built

Construction of the Umiam Hydroelectric Project formally began on January 9, 1960, when Prime Minister Nehru stopped at the project site on a three-day tour of Assam. Although he arrived 90 minutes behind schedule, he was greeted by a large crowd awaiting a glimpse of their Prime Minister. Disdaining security precautions taken by the state police, Nehru crossed a wooden fence to greet the assembled masses. To inaugurate the Umiam Project, he threw a switch that set off the first explosive charges of the project. The explosion blasted the first hole for the river diversion tunnel and sent rocks flying several hundred feet.[25]

The diversion tunnel, once completed, shunted the Umiam River around the construction site of the main concrete dam. The dam was built in a narrow ravine where the river took a sharp turn from the north to the east. Cofferdams on either side of the construction site prevented river water from flooding the dam as it was being built. (The remains of the downstream cofferdams are still visible at the foot of the dam. From the roadway that crosses the crest of the main dam, the cofferdams look like little concrete boxes.)

The first step for constructing the main dam was preparing the foundation. Thousands of laborers worked alongside mechanical excavators to strip the foundation to bedrock and pry out loose rocks with hammers and chisels. A total of 3.9 million cubic feet of earth and rock had to be removed before the pouring of the foundation could begin.[26]

The labor force on the project included local labor as well as skilled laborers brought in from other parts of the country. ASEB housed the laborers in two purpose-built colonies—a large one at the concrete dam

Rs 4 million in compensation. "Power Projects in Assam; Electricity Board Praised," *AT*, March 31, 1961.

[25] "Nehru Envisages Assam's Great Future; Umiam Hydel Project Inaugurated," *AT*, January 10, 1960; "Umiam Hydel Project," *AT*, January 9, 1960; Jawaharlal Nehru, speech at public meeting in Shillong, January 9, 1960, in Madhavan K. Palat, ed., *Selected Works of Jawaharlal Nehru: Second Series* (New Delhi: Jawaharlal Nehru Memorial Fund, 2014), 56:232–39. The next day, Nehru laid the foundation stone of the Brahmaputra Bridge at Pandu (see Chapter 6).

[26] Robert Gilkey, "Assam's 'Weeping River' Power Project," *AT*, June 18, 1961; "Progress of Umiam Hydel Project: Commission Expected Next Year," *AT*, May 16, 1962; Assam State Electricity Board, "Umiam Hydro-Electric Project," commemorative pamphlet, n.d. [1966], Office of Director (Generation), Meghalaya Energy Corporation Ltd., Shillong.

site and a smaller one near the powerhouse. In addition to common hazards of heights and blasting faced by workers at any construction site, the laborers at the Umiam Project faced further dangers in the Khasi forests. In December 1960, a tiger mauled three project workers, one seriously. The local authorities employed a *shikari* (big-game hunter) and some soldiers to track down and kill the tiger.[27]

In a construction project that was, as usual, full of ceremonies marking important milestones, the first concrete for the project was poured on April 16, 1962. Assam state governor S.M. Shriganesh presided over the ceremony, after which he was presented with a commemorative silver model of a concrete bucket. In the ensuing months, load after load of concrete was deposited into the body of the dam, 7.7 million cubic feet in all. Consisting of cement from Bihar and aggregate from a quarry near the construction site, the concrete was poured from a bucket lift that traveled on a cableway extending across the ravine.[28]

Bucketload by bucketload, the dam grew to a design prepared by the Central Water and Power Commission in Poona. Formed in 1948 by the amalgamation of two institutions established in the late colonial period, CWPC had the mandate to initiate and support, in cooperation with state governments, engineering projects for the use of national water resources for irrigation, navigation, electric power generation, and flood control. The Hydro-Electric Directorate of the CWPC Power Wing was responsible for the design of Umiam Dam and other hydroelectric projects built across India during the early independence period[29] (Fig. 7.2).

CWPC's laboratory, the Central Water and Power Research Station (CWPRS), had facilities to support full hydrodynamic model studies of dams, canals, ports, and bridges. These and other facilities were put to good use for the design of Umiam Dam. On this project, site geology required that part of the foundation would have to be built on sloped rock. This created particular challenges in the analysis of the structure. To analyze the stresses that would develop in the dam because of the sloped foundation, CWPRS employed isochromatics, in which a scaled cross-section of the dam was cut out of plastic and placed in a rig that

[27] "Tiger Mauls Three Umiam Employees," *AT*, December 12, 1960.

[28] ASEB, "Umiam Hydro-Electric Project," commemorative pamphlet.

[29] K.C. Khatri, "Organisational Set-Up of the Central Water and Power Commission," *IJPRVD* 9, no. 9 (September 1959): 45–49.

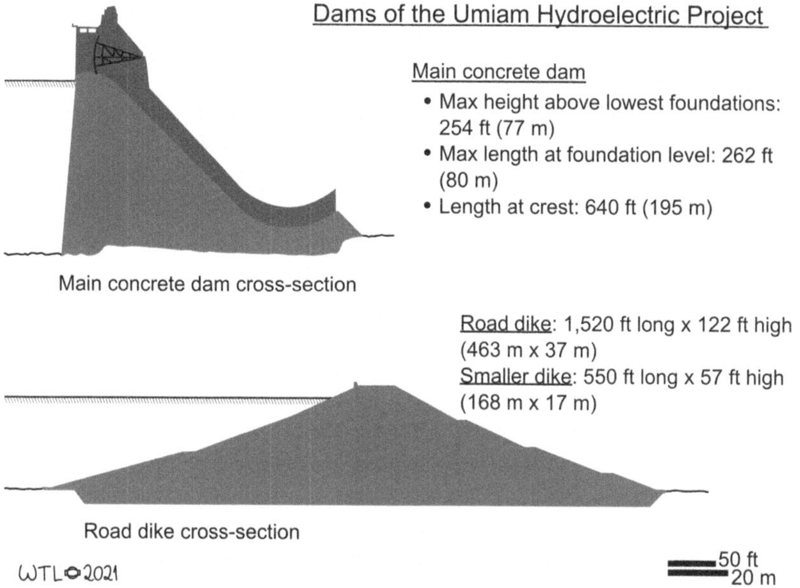

Fig. 7.2 The dams of Umiam Hydroelectric Project (Drawing by the author, based on pictures in Assam State Electricity Board, "Umiam Hydro-Electric Project," commemorative pamphlet, n.d. [1966], Office of Director [Generation], Meghalaya Energy Corporation Ltd., Shillong)

put pressure on the test article to simulate the loading of the dam. In the test article, the loads produced tiny deformations in the plastic cross-section, which appeared as rainbow patterns when viewed through cross-polarized light. The patterns allowed the engineers at CWPRS to visualize the stresses developed inside the dam without employing long and costly numerical calculation methods.[30]

CWPC was quite capable of designing Umiam Dam, as it already had over a decade's worth of experience on similar projects. ASEB had never

[30] C.L. Handa and B. Pant, "Role of Experimental Methods in the Solution of Foundation Problems," *Journal of the Indian National Society of Mechanics and Foundation Engineering* 4 (January 1965): 18–21; "Introduction with Vision and Mission," *Central Water and Power Research Station*, http://cwprs.gov.in/Page/Introduction.aspx (accessed November 8, 2015).

built anything like Umiam Dam before, but it too was capable of implementing the project. After the US government extended full financing to the project, DLF tried to compel ASEB to hire foreign consultants to oversee design and construction. Wary of the costs and confident in their own abilities, ASEB resisted, and DLF ultimately relented.[31]

ASEB did hire one American consultant for the project, Professor Roy W. Carlson of the University of California–Berkeley. Since 1925, this MIT-educated professor had worked on dams in the United States and abroad. He also ran a consulting business for supplying and installing instrumentation to monitor dams. ASEB ordered $16,000 worth of instrumentation from the professor's catalog, including resistance thermometers, strain meters, stress meters, and pore pressure cells. Carlson traveled to Assam and oversaw his equipment's installation in the dam.[32]

At the northern end of the Umiam Valley, Bahri & Co., a private concern, built two earthen dikes, which plugged other potential escape routes for reservoir water. (Professor Carlson installed instrumentation in these dikes as well.) To build the dikes, Bahri & Co. used imported heavy earthmoving machinery, Caterpillar D 8 tractors and DW 21 scrapers. Caterpillar machinery had recently been made available on the Indian market by the authorized dealer Tractors (India) Ltd. The dealer staged programs in which its machines demonstrated earthmoving capabilities in front of potential buyers and their employees. Bahri & Co.'s Caterpillar machines had already seen a fair amount of use and needed to be overhauled with spare parts. DLF granted permission for the foreign-exchange loan to be used for the parts.[33]

[31] Sundarajan to C.S. Krishna Moorthi, November 9, 1960, file no. 15(11)60, Ministry of Finance, NAI; Agency for International Development, *Program and Project Data Related to Proposed Programs—FY 1967: Near East and South Asia* (Washington, DC: n.p., 1966), 71.

[32] Acceptance of tender, India Supply Mission to Roy W. Carlson, August 8, 1961, file no. 15(11)60, Ministry of Finance, NAI; Assam State Electricity Board, "Umiam Hydro-Electric Project," commemorative pamphlet, n.d. [1966], Office of Director (Generation), Meghalaya Energy Corporation Ltd., Shillong. A seismograph was also installed at the dam site. A.N. Tandon and H.M. Chaudhury, "A Report on the Seismicity Studies by the Indian Meteorological Department," *Symposium on Earthquake Engineering* (1966): 34–43; J.M. Raphael and R.W. Carlson, *Measurement of Structural Action in Dams* (Berkeley, CA: James J. Gillick & Co., 1965), WRCA.

[33] "Technical Notes and News: New Caterpillar Equipment," *IJPRVD* 11, no. 2 (February 1961): 31–32; K.P.S. Nair, memo, December 5, 1960, file no. 15(11)60,

Further to the north, on the other side of a hill and at a lower elevation, ASEB built the Stage I powerhouse for the project. Work crews bored a 7,000-foot tunnel directly through the hill from an intake in the reservoir to a point on the hillside above the powerhouse. Once the project was complete, the tunnel would carry water from the reservoir to the penstocks, two 1,800-foot steel-lined tubes that sloped down the hill to feed the turbine generators in the powerhouse. Since the powerhouse was at a lower elevation than the lake, the water would have higher head (pressure) when it reached the turbines, thus generating more electricity than if the powerhouse had been located at the foot of the dam.[34]

After water passed through the turbines and exited the Umiam powerhouse by the tailrace, it flowed into Sumer Stream, a tributary of the Umtru River. The water was diverted into the Umtru River so that the Meghalaya Plateau's first hydroelectric project could have a reliable supply of water from the Umiam Reservoir. Umtru was a run-of-river project, which raised the level of the reservoir just enough to produce sufficient head to turn the turbines, but not enough to create a proper storage reservoir. The rivers of the Khasi Hills are fed entirely by rainfall. During the rainy season months, from April to October, the rivers run high while draining some of the heaviest rainfall in the country. But during the dry season, with no rain to feed them, river levels sink considerably as aquifers drain. With its capacity to store water, Umiam Reservoir provided a consistent water supply for both the Umiam and Umtru hydroelectric projects.[35]

To carry the electricity generated by the Umiam Project to the places where it might be used, ASEB built transmission lines, thus creating Assam's first interconnected power grid. Before the construction of the Umiam Project, generation and use of electricity in Assam was localized, with diesel generating stations providing power directly to the cities and

Ministry of Finance, NAI; M. Gopalakrishnan, memo, January 10, 1961, ibid.; K.V. Rajagopalan to DLF Loan Services Staff, February 21, 1961, ibid.

[34] Assam State Electricity Board, "Umiam Hydro-Electric Project," commemorative pamphlet, n.d. [1966], Office of Director (Generation), Meghalaya Energy Corporation Ltd., Shillong; "Umiam Hydel Project," *AT*, January 9, 1960.

[35] "Canada and the Colombo Plan: The Umtru Project," *External Affairs* 9 (August 1957), 241–43; "Umiam Hydel Project," *AT*, January 9, 1960; "Umiam Hydel Project," *Indian Railway Gazette* 62, no. 9 (September 1964), 239.

towns they served. The only major power station was Umtru Hydroelectric Project, which served Gauhati through a short transmission line. As part of the Umiam Project, ASEB built transmission lines that stretched across lower Assam. Major lines with a voltage of 132 kV ran from the powerhouse at Sumer up to Shillong and down to Gauhati via Umtru powerhouse. Lesser lines of 33 kV extended the network from Shillong to Cherrapunjee on the southern side of the Meghalaya Plateau overlooking East Pakistan. In later expansions of the transmission network, lines reached to Namrup in upper Assam, where a 69-MW gas-turbine generating station was built concurrently with the Umiam Project. Wires strung from towers built on ridges upstream of Saraighat Bridge stretched 6,295 feet across the Brahmaputra, carrying power to the towns on the north bank of the river.[36]

The Umiam reservoir flooded part of the original route of the Gauhati–Shillong Road. The highway had to be rerouted higher up on the valley wall. It crossed the Umiam gorge by running across the crest of the dam itself. This new section of road had two lanes, one for each direction. While the highway was being rerouted, the Assam Public Works Department set out to widen the rest of the highway to two-lane width. Funds for road construction were tight, though, and it was not until the Fourth Five-Year Plan that the PWD was able to finish the two-laning work and abolish the system of gate timings on the highway.[37]

The expansion of the highway was financed by the government's own tax base. The construction of the dam, powerhouse, and transmission lines, though, was financed by the government of the United States—an arrangement that had its own challenges, pitfalls, and rewards.

[36] Citation: M.R. Sitaram, "Report on the Transmission Line Umiam (Barapani) Project," n.d. [April 1961], file no. 15(11)60, Ministry of Finance, NAI; K.P.S. Nair to C.L. Garg, July 31, 1961, ibid.; "Break Down Cost of Equipment and Labour per Mile of ... Double Circuit Line..." n.d., ibid.; "Now...to Umiam Project," *Andhra Pradesh*, August 1965, 33; Assam State Electricity Board, "Umiam Hydro-Electric Project," commemorative pamphlet, n.d. [1966], Office of Director (Generation), Meghalaya Energy Corporation Ltd., Shillong; A.K. Ganguly, "Major Power Stations in Assam," *IJPRVD* 19, no. 8 (August 1969): 342–44.

[37] "Progress of Umiam Hydel Project: Commission Expected Next Year," *AT*, May 16, 1962; Government of Assam Public Works Department, "Report on the Administration of Public Works Department (Roads & Buildings Wing), Assam for the Year 1963–64," Assam State Archives; Raghuraj Singh, "Additional Allocation for the Third Five Year Plan to Assam Government," memo, August 2, 1962, file no. WI-33(6)/62, Ministry of Transport, NAI.

How the Dam Was Financed

Expenditures on major development projects like Umiam Dam would fall into one of two categories: domestic costs and foreign exchange. Domestic costs were payable in rupees, as they were not spent outside of the country, but foreign exchange usually had to be used for importing anything from abroad, as the Indian rupee was not considered hard currency on the international market. The US government accordingly offered two loans for the Umiam Project. The first was DLF loan no. 141 (later redesignated USAID loan 386-A-030), which provided up to $2.5 million of foreign exchange, with a twenty-year repayment period at 3.5 percent interest. The other was a PL-480 loan of Rs 64 million ($13.1 million in US currency at the time), meant to cover domestic costs of the project. Both were repayable in rupees.[38]

The Development Loan Fund had first provided financing for Indian electrification in 1958 with a general development loan of $10 million. DLF Loan no. 141—announced in July 1960 and signed on December 5—was part of an expansion of US involvement in Indian power development. In total, eight Indian power projects received DLF foreign-exchange financing in 1959–1960, with a combined value of $78.6 million. Up to $50 million dollars between the eight loans could be spent outside the United States, although future DLF loans would include a strict American-purchases-only policy. The money could not be spent in eastern-bloc countries, and under the "small business notification" requirement, American businesses needed to get the first chance to fill contracts.[39]

[38] Development Loan Fund, Loan No. 141, Loan Agreement, December 5, 1960, file no. 15(11)60, Ministry of Finance, NAI; Telegram, C.S. Krishna Moorthi to N.C. Sengupta, November 5, 1960, ibid.; House Committee on Appropriations, Foreign Appropriations for 1963: Hearing before a Subcommittee of the Committee on Appropriations, 87th Cong., 2nd sess., August 1, 1962, 403; Agency for International Development, *Program and Project Data Related to Proposed Programs—FY 1965: Near East and South Asia* (Washington, DC: n.p., 1964), 84.

[39] United States Department of Commerce, *Status of Active Foreign Credits of the United States Government and of International Organizations as of June 30, 1966* (Washington, DC: Government Printing Office, 1966), 55; Barapani Hydel Project summary sheet, attachment to letter, Harold W. Feldt to Ralph Trisko, February 11, 1960, box 2, USAID Mission to India Subject Files 1955-62, RG 286, NARA; "Six Proposed DLF Power Projects (GOI Submission of June 2, 1959)," memo, ibid.; House Committee on Appropriations, Foreign Appropriations for 1963: Hearing before a Subcommittee of the Committee on Appropriations, 87th Cong., 2nd sess., August 1, 1962, 403. The loans were for six thermal plants located in the cities of Ahmedabad (extension of an existing

For Umiam Dam, ASEB ordered construction equipment from the United States. A ten-ton movable crane was sold to the project by Rocky Mountain Export Co. of Colorado. In addition to the used tractors and scrapers brought to the project by Bahri & Co., Caterpillar supplied several new machines for the construction, including three D 4 tractors, two D 8 tractors, and three 619 scrapers. Other American-sourced equipment included drilling sets and an ammonia printing and developing machine for making blueprints.[40]

Most of the imported machinery and equipment actually installed in the project came from third countries, as the purchases fell under the $50 million limit. Turbine generators and 132 kV insulators came from Japan. Companies in Britain supplied testing equipment and cables for the transmission lines. A testing generator and transformer came from Austria. Germany supplied DC ammeters, France transmission line erection tools, and Sweden switchgear.[41]

After the signing of the formal agreement for DLF Loan no. 141, the Indian government did not automatically receive a lump sum of $2.5 million for the Umiam Project. In fact, the government got no money at all—and would not get any until DLF had received receipts and other documentation for money that had already been spent for the project. The original agreement stated what kinds of purchases the loan would and would not cover, but the wording of the agreement left some ambiguity about certain purchases. On these points of ambiguity, the Ministry of Finance and Central Water and Power Commission would continue to renegotiate the terms of the loan throughout project implementation. As seen in the opening story about the Czechoslovakian diesel generators, Finance and CWPC served as intermediaries between the US government and the Assam State Electricity Board. As we will see shortly, the Indian Embassy in Washington also served as intermediary and gatekeeper.

plant), Barauni, Birsinghpur, Chandrapura, Durgapur (extension), and Kanpur (extension); as well as two hydroelectric projects, Sharavathi in Mysore State (Karnataka) and Umiam.

[40] T.V. Thadani to C.L. Garg, August 28, 1962, file no. 15(11)60, Ministry of Finance, NAI; Daya Shankar to ASEB Chief Engineer (Hydro-Electric), May 4, 1961, ibid.

[41] T.V. Thadani to C.L. Garg, August 28, 1962, file no. 15(11)60, Ministry of Finance, NAI; Daya Shankar to ASEB Chief Engineer (Hydro Electric), August 7, 1961, ibid.

The Indian government had to spend foreign exchange before it could receive the loan funds, but how could it spend money it did not have? The fact was that the government did have some foreign exchange, saved up from exporting Indian goods. The foreign-exchange reserves were not enough to finance all the projects the government was undertaking at once, but they could be doled out to cover expenses until loan funds were disbursed. When the US government disbursed loan funds, the money went back into the Indian government's foreign-exchange reserves, and could then be used for other purposes.[42]

Although the original financing agreement called for the construction of "a 27,000 KW hydroelectric plant on the Umiam River in the State of Assam" with "related transmission lines and two substations," ASEB remained undecided about whether it wanted to build a station with three or four 9-MW turbine generators. When ASEB issued a tender notice for the turbine generators, it requested bids for three 9-MW units with the possible addition of a fourth to the order. In response, ASEB received fifteen bids from manufacturers in seven different countries, none of which were based in the United States. ASEB decided that the offer of Japanese firm Toshiba best met its needs. After receiving the required clearance from DLF for offshore procurement (i.e., procurement from a non-US supplier), the electricity board placed an order for three turbine generators from Japan. In conjunction with the order, three ASEB engineers traveled to Tokyo to receive training from the manufacturer for operation of the turbine generators.[43]

After placing the order with Toshiba, ASEB decided that it wanted a fourth turbine generator after all. Toshiba already knew that this was a possibility, and the Planning Commission had agreed to place the fourth unit in the Third Five-Year Plan. But there was difficulty with the financing. As this was an additional offshore purchase, it would have to be approved by DLF. The Americans would have preferred that the fourth unit be purchased from a US supplier, as the $50 million cap for offshore

[42] Development Loan Fund, Loan No. 141, Loan Agreement, December 5, 1960, file no. 15(11)60, Ministry of Finance, NAI.

[43] Development Loan Fund, Loan No. 141, Loan Agreement, December 5, 1960, file no. 15(11)60, Ministry of Finance, NAI.; B.C. Kapur to G.R. Iyer, June 23, 1960, ibid.; telegram, N.C. Sengupta to C.S. Krishnamoorthi, August 22, 1960, ibid.; M.R. Sita Ram to Finance Ministry attn. K.R. Chari, July 7, 1961, ibid.; T.K. Bandopadhyay, "Umiam Hydro-Electric Project: Efficiency Tests for Turbines," *IJPRVD* 17, no. 5 (May 1967): 13–16, 25; "Barapani Hydel Project: Jap Firm to Supply Turbines," *AT*, May 16, 1961.

purchases was about to be reached. But since no American suppliers had made bids for the first three units, it seemed unlikely that a second round of tenders would be any more fruitful than the first.[44]

In early 1961, C.S. Krishnamoorthi and P. Govindan Nair of the Indian embassy in Washington asked DLF about the possibility of using the loan to finance the fourth turbine generator. John Ulinski, the regional director for South Asia, indicated that the question would have to be cleared by the board first, and before he could present the question, he would need some more information, including project details and the justification for the additional unit. In response, ASEB wrote up a draft report with the required information. The draft made it as far as the Indian embassy in Washington before A.K. Ghosh returned it to the Finance Ministry. "I started to modify the language of the note in order to correct the typing mistakes," he said in an attached note, "but while doing so, I found I could not understand the justification given for the fourth generating set." The staff of the Power Wing of the Central Water and Power Commission produced a revised project description, which the Indian embassy relayed to DLF in Washington. The American agency again asked for further information, this time about offshore purchases already made for the Umiam Project.[45]

By the end of 1961, it was clear to embassy staff in Washington that DLF was opposed to financing a fourth unit purchased from Japan. Krishnamoorthi reported in a letter to the Finance Ministry that the subject should not be taken up before February of the next year. In February, a telegram from the Finance Ministry to the embassy reported that the decision had been made in New Delhi to drop the matter of financing the fourth unit from the American loan altogether. The Government of India, realizing the direction the negotiations were heading, decided to withdraw of its own volition, rather than waiting for the American government to give them the inevitable rejection. This was what would happen

[44] P.P. Agarwal to K.S. Sundara Rajan, December 15, 1960, file no. 6(16)/61 vol. 3, Ministry of Finance, NAI; N.C. Sen Gupta to C.S. Krishna Moorthi, December 23, 1960, ibid.

[45] P. Govindan Nair to N.C. Sen Gupta, January 4, 1961, file no. 6(16)/61 vol. 3, Ministry of Finance, NAI; ASEB, "Information Furnished to Foreign Exchange Cell (CW & PC)," n.d. [July 1961], ibid.; A.K. Ghosh to C.S. Swaminathan, August 14, 1961, ibid.; T.V. Thadani to C.L. Garg, August 28, 1961, ibid.; "Barapani Hydro-Electric Project (Assam): Installation of IV Generating Unit," n.d. [August 1961], ibid.; C.S. Swaminathan to T.V. Thadani, October 31, 1961, ibid.

on a much larger scale with an aid request for Bokaro Steel Plant the next year. These carefully planned diplomatic maneuvers were meant to keep the providers of aid happy and to save face on both sides. Ultimately, ASEB placed an order with Toshiba for the fourth unit, with the Japanese government providing ad hoc financing.[46]

Under the terms of the original agreement for DLF Loan no. 141, the US government would disburse loan funds only for documentation received by December 31, 1964, unless the aid agency agreed to extend the deadline. By request of the Indian authorities, the deadline was extended to the end of 1965, and then further to the end of 1966. For the first extension, ASEB claimed that it was still making incremental payments for machinery already installed at the project site, and was also paying wages to foreign specialists overseeing the installation of their companies' equipment. For the second extension, ASEB wanted to replace a shipment of 33 kV circuit-breaker kiosks from West Germany, which had been impounded while passing through East Pakistan on the Brahmaputra River at the outbreak of the war between India and Pakistan in September 1965. When the second extension expired, ASEB applied for yet another extension, in hopes of reordering the impounded kiosks. USAID rejected this third extension request, because the project had already been producing electricity commercially for more than a year, so there was no clear need for the missing kiosks.[47]

By the time the last disbursement deadline passed, USAID had only provided $1.9 million of the maximum authorized $2.5 million for the Umiam Project. ASEB had spent an additional $90,000 in foreign exchange that USAID refused to cover under the loan. The aid agency accordingly deobligated $641,000 of the loan. Based on a new amortization schedule, the Indian government was slated to pay a total of $2.3

[46] Telegram, C.S. Krishnamoorthi to K.S. Sundara Rajan, December 23, 1961, file no. 6(16)/61 vol. 3, Ministry of Finance, NAI; telegram, C.S. Swaminathan to A.K. Ghosh, February 19, 1962, ibid.; Assam State Electricity Board, "Umiam Hydro-Electric Project," commemorative pamphlet, n.d. [1966], Office of Director (Generation), Meghalaya Energy Corporation Ltd., Shillong.

[47] Ibid.; "Format to Accompany Borrower's Request for Extension of Terminal Date for Disbursement," DLF Loan 141, June 27, 1964, file no. 15(11)60, Ministry of Finance, NAI; Barapani project summary, attachment to letter, M.R. Sita Ram to C.L. Garg, December 9, 1964, ibid.; John A. Ulinski to GOI Ministry of Finance, December 15, 1964, ibid.; N. Vishwanathan, memo, April 10, 1967, ibid.

Fig. 7.3 Aerial view of Umiam Lake in 2010 (Photo by the author)

million ($400,000 of which was interest) by the last payment date of May 7, 1977.[48]

Contrary to the intent of the original 1960 loan agreements, the US government did not, in the end, finance the Umiam Project in full. As the project was being built, the United States could not—or would not—finance everything. At least in India, development aid was not largesse handed out on the understanding that it would be used in the best way. The United States was particular about the way its aid was used—so particular, in fact, that full financing was made impossible by restraints placed on the project by DLF and USAID. Lacking full American financing, the project had to be completed with financing from Japan and the Indian government's own resources (Fig. 7.3).

[48] Implementation Letter No. 11, Thomas L. Cranmer to R.K. Jerath, June 19, 1967, file no. 15(11)60, Ministry of Finance, NAI.

Conclusion

On September 11, 1965, the fourth and last turbine in the powerhouse of Umiam Stage I came online and began feeding power into the Assam electrical grid. It was a moment long in the making. Stage I had originally been scheduled for completion in early 1963, but the construction schedule kept slipping further and further behind. The expansion of the project from 27 MW to 36 MW slowed down construction, as did delays in receiving certain key components. It is also likely that ASEB's inexperience with building large hydro projects delayed construction as well, as the utility was learning as it went along. In any event, the last turbine came online two and one-half years behind the originally planned commissioning date. The new project was dedicated early the following year. The President of India, Dr. Sarvepalli Radhakrishnan, had originally planned to attend the dedication, but in the end he had to bow out because his physicians had told him that his poor health at the time made a trip to Assam inadvisable.[49]

Throughout planning and construction of the Umiam Project, ASEB had touted the economic benefits that the power generated in the Stage I powerhouse would bring to Assam. A report printed in *Assam Tribune* in January 1960 claimed: "the power output of the project will adequately meet the power requirements of industries such as the Cement Factory at Cherrapunji, the Ropeway between Cherrapunji and Pandu, the Spun Silk Mill at Jagi Road, Spinning Mill, Oil Refinery, Jute Mill, Cycle Factory, Re-rolling Mill etc. and a number of other industries which have been approved by the Planning Commission and whose power requirements could not be met from the original plan schemes." The report noted that power would also be used for rural electrification and agriculture, "and thus help in improving the economic condition of the people in areas served by it."[50]

At first, though, the project offered too much power. Along with the Namrup gas plant, the Umtru project, and diesel generating sets, ASEB was able to generate 160 MW of power by 1965. An article in *Indian Engineer* in March of that year reported that the electricity board had found buyers for only 60 MW, thus leaving more than half of the grid's

[49] "President to Visit Assam," *AT*, January 25, 1966; "President's Assam Visit Put Off," *AT*, February 20, 1966.

[50] "State Electricity Board and Its Work," *AT*, January 9, 1960.

capacity unsold. Despite the long litany of industries named in the 1960 report, the anticipated industrial investment was slow to materialize. The *Indian Engineer* article attributed sluggish investment to fears of further Chinese incursions into northeast India. B.C. Kapur, chair of the electricity board, told the press that he expected industrial investment and demand for power to pick up soon.[51]

Kapur was more than proven correct over the following decades, when rising demands for power shot past the state electricity board's capacity for a generation. When Meghalaya State Electricity Board split off from ASEB in 1975 (three years after the Meghalaya Plateau became its own state), the utility generated surplus power that it could sell to neighboring utilities. At this time, the utility was building additional dams and powerhouses to extract more power from the water of the Umiam River as it flowed down toward Umtru powerhouse. When Umiam Hydroelectric Project Stage IV came online in 1992, the combined Umiam–Umtru Project had a potential generative capacity of 185.2 MW. But it would not be long before industrial development, population growth, and the escalating wants of the privileged classes would begin to create demands for power that the state grid was unable to meet. Meghalaya Energy Corporation Ltd., the public-sector electric utility that now operates the Umiam Project, has to pay heavily to buy electricity from the North Eastern and national power grids. But even this is insufficient. In the summer months, when demand for power is high and the Umiam reservoir is low, the utility must cut service to selected areas. These scheduled blackouts, known as load shedding, have become a fact of life in areas served by power from the Umiam Project.[52]

[51] "Barapani Hydro-Electric Project's Progress," *Indian Engineer*, March 1965, 34–37.

[52] Meghalaya Energy Corporation Limited, "Existing Power Stations," http://meseb.nic.in/network.htm (accessed April 18, 2012); Agency for International Development, *Program and Project Data Related to Proposed Programs—FY 1967: Near East and South Asia* (Washington, DC: n.p., 1966), 71; Uttama Choudhury, "River Umiam and Its Role in Hydro-Electricity Generation, Tourism, and Industrial Growth," in *Rivers and Riverine Landscape in North East India*, ed. S. Sengupta et al. (New Delhi: Concept Publishing, 2006), 136, 138; Meghalaya Energy Corporation Limited, "Stage IV Power Station," http://meseb.nic.in/psstage-iv.htm (accessed April 18, 2012); "Demand of Power in Meghalaya," *Meghalaya Energy Corporation Limited*, http://meecl.nic.in/demand.htm (accessed November 6, 2012); "No Immediate Load Shedding in State: MeECL," *Shillong Times*, October 8, 2012.

The Umiam Hydroelectric Project was one of many attempts undertaken in India during the early independence period to engineer the landscape and bring a part of the country into the modern industrial age by bestowing upon it a plentiful supply of electricity. Like probably every such project, Umiam Dam was only able to deliver on its promises in part. The project's powerhouse generated electricity for the cities of Assam, and the reservoir solved the problem of the seasonal variability of the rivers of the Meghalaya Plateau by storing water through the dry season. The lake also provided a place for recreation and a scenic backdrop for the drive up to Shillong. But the lake is increasingly polluted, as the streams of Shillong drain into the lake—and Shillong has no sanitary sewer. It was not until the early 1980s that the first environmental study of the lake was commissioned, and this report expressed ambivalence over whether the environment of the lake should be protected at all, because the reservoir had originally been built solely for the purpose of hydroelectricity. But the fact is that Umiam Lake is a part of the environment, whether the planners intended it that way or not. The reservoir was built in a period that Ramachandra Guha calls an "age of ecological innocence," but that age has passed. More recently, various levels of government, academics, and private conservation groups have begun to study and try to care for the environment of Umiam Lake.[53]

Umiam Dam is an example of a development project making use of resources initially belonging to indigenous minorities for the benefit of the nonindigenous majority population. The Khasi landowners who had originally occupied the Umiam Valley received compensation for the loss of their land, but it is unlikely that the landowners could ever have prevented the sale. In Indian development during the early independence period, the government almost always got its way. Although popular movements against development later in the twentieth century placed a check on the government's power, the pattern of exploitation seen in Umiam Dam persists. Throughout India, and around the world,

[53] Central Board for the Prevention and Control of Water Pollution, *Environmental Status: Barapani Lake, Meghalaya* (New Delhi, n.d. [1982–83]), NLI; Amitangshu Acharya, "Umiam, Shillong's 'Wasted' Pride," *Calcutta Telegraph*, September 30, 2009; Ramachandra Guha, *How Much Should a Person Consume? Thinking Through the Environment* (Gurgaon: Hachette India, 2010), 45; Bashida Massar, Sudip Dey, and K. Dutta, "An Electron Microscopic Analysis on the Ultra Structural Abnormalities in Sperm of the Common Carp *Cyprinus carpio L.* Inhabiting a Polluted Lake, Umiam (Meghalaya, India)," *Microscopy Research and Technique* 74, no. 11 (April 2011): 998–1005.

socially marginal indigenous people continue to be exploited because they inhabit lands with resources that the majority population craves.[54]

As ASEB acquired land in the indigenous borderlands of Assam, it also turned to the United States, the world's richest country, for financing the project it would build there. Getting financing from the US government was a complex political arrangement that entailed balancing the needs of the recipient with the sensibilities and regulations of the donor. The Finance Ministry of the central government served as mediator and gatekeeper for all interactions between the opposite parties of the loan agreement. In this role, Finance was able to avert more than one diplomatic faux pas.

The Umiam Hydroelectric Project represents a transitional stage in the industrialization of India. During construction of the first stage of the project in the 1960s, the institutions for designing hydroelectric plants were already established in India, and ASEB only needed to bring in limited foreign technical assistance for the peculiar problems of building and monitoring a dam in a seismically active area. But even though expertise was present in India, equipment (with a few exceptions) and capital were not. By using foreign financing, the project was able to import construction machinery, turbine generators, transformers, and other specialized equipment. But from beginning to end, Umiam Stage I was designed and implemented by Indians.

In several respects, Umiam Dam was a typical development project of early independent India. Its construction entailed a mixture of foreign and domestic inputs, it was touted as a vector of modernization for a backward area, and it used tribal lands to benefit the non-tribal majority. The project drew on long traditions, both foreign and domestic. There was Indian hydraulic engineering and dam-building, which dated back to ancient times. Then there was the technology of hydroelectric power generation. Although hydroelectricity itself was first generated only in the nineteenth century, the use of rivers to generate mechanical hydropower dated back as far as the Roman Empire and Han China.

[54] For state exploitation of resources on land inhabited by indigenous or otherwise marginal people, see Peter Penz, "Dams, Guns, and Refugees," *Alternatives Journal* 30, no. 4 (September/October 2004): 8–21; Meeta Deka, "Mega-Dams, Environment and History in North-East India," *Dialogue: A Quarterly Journal of Astha Bharati* 13, no. 4 (April–June 2012), http://www.asthabharati.org/Dia_Apr%20012/me.htm; Nick Estes, *Our History Is the Future: Standing Rock Versus the Dakota Access Pipeline, and the Long Tradition of Indigenous Resistance* (London: Verso Colophon, 2019).

For the last two chapters of this book, we will turn our attention to a rather more unconventional development project: Tarapur Atomic Power Station, the first commercial nuclear plant built in India. Unlike hydropower, the principles of deriving power from the splitting of the atom had only been understood for a matter of decades, not centuries, by the 1960s. Nuclear power was something new under the sun. As we will see, India did have atomic research laboratories, but the domestic knowledge base was still in its infancy and the plant had to be designed abroad and built by foreign contractors. Most significantly, the plant remained linked with foreign countries after its completion because its fuel could not be obtained domestically. The Umiam Project was built to use water that fell for free in the Khasi Hills; Tarapur would use uranium that had been enriched in a reactor in the West and shipped to India under international safeguards.

References

1. Agency for International Development. *Program and Project Data Related to Proposed Programs—FY 1965: Near East and South Asia*. Washington, DC: n.p., 1964.
2. Agency for International Development. *Program and Project Data Related to Proposed Programs—FY 1966: Near East and South Asia*. Washington, DC: n.p., 1965.
3. Agency for International Development. *Program and Project Data Related to Proposed Programs—FY 1967: Near East and South Asia*. Washington, DC: n.p., 1966.
4. Bandopadhyay, T.K. “Umiam Hydro-Electric Project: Efficiency Tests for Turbines.” *IJPRVD* 17 (May 1967): 13–16, 25.
5. Bangash, Yaqoob Khan. *A Princely Affair: The Accession and Integration of the Princely States of Pakistan, 1947–1955*. Karachi: Oxford University Press, 2015.
6. Bareh, Hamlet Ngap Kynta, *The Economy of Meghalaya: Tradition to Transition*. Guwahati: Spectrum Publications, 2001.
7. Bareh, Hamlet. *The History and Culture of the Khasi People*. Calcutta: Naba Mudran, 1967.
8. “Canada and the Colombo Plan: The Umtru Project.” *External Affairs* 9 (August 1957), 241–43.
9. Central Water Commission. “National Register of Large Dams—2009.” http://www.cwc.nic.in/main/downloads/National%20Register%20of%20Large%20Dams%202009.pdf (accessed October 25, 2012).
10. Das, J.N. *A Study of the Land System of Meghalaya*. Guwahati: Law Research Institute of Gauhati High Court, 1990.

11. Desai, Padma. *The Bokaro Steel Plant: A Study of Soviet Economic Assistance.* New York: American Elsevier Publishing Company, 1972.
12. Ganguly, A.K. "Major Power Stations in Assam." *IJPRVD* 19, no. 8 (August 1969): 342–344.
13. Gopal, Sarvepalli, and Madhavan K. Palat, eds. *Selected Works of Jawaharlal Nehru: Second Series.* New Delhi: Jawaharlal Nehru Memorial Fund, 1984–2014.
14. Guha, Ramachandra. *How Much Should a Person Consume? Thinking Through the Environment.* Gurgaon: Hachette India, 2010.
15. Handa, C.L., and B. Pant. "Role of Experimental Methods in the Solution of Foundation Problems." *Journal of the Indian National Society of Mechanics and Foundation Engineering* 4 (January 1965): 3–33.
16. "India–Canada Co-operation in River Valley Development." *IJPRVD* 12, no. 10 (October 1962): 19–20, 26.
17. Khatri, K.C. "Organisational Set-Up of the Central Water and Power Commission." *IJPRVD* 9, no. 9 (September 1959): 45–49.
18. Klingensmith, Daniel. *"One Valley and a Thousand": Dams, Nationalism, and Development.* New Delhi: Oxford University Press, 2007.
19. *Large Dams in India.* Vol. I. New Delhi: Central Board of Irrigation and Power, 1987.
20. Moraes, Frank, ed. *The Indian and Pakistan Year Book and Who's Who 1951.* Bombay: Times of India, 1951.
21. Penz, Peter. "Dams, Guns, and Refugees." *Alternatives Journal* 30, no. 4 (September/October 2004): 8–21.
22. Singh, Satyajit. *Taming the Waters: The Political Economy of Large Dams in India.* Delhi: Oxford University Press, 1997.
23. "Starting Production of Heavy Power Equipment in India." *Czechoslovak Heavy Industry*, May 1966, 20.
24. Tandon, A.N., and H.M. Chaudhury. "A Report on the Seismicity Studies by the Indian Meteorological Department." *Symposium on Earthquake Engineering* (1966): 34–43.
25. "Technical Notes and News: New Caterpillar Equipment." *IPRVD* 11, no. 2 (February 1961): 31–32.
26. Tod, James. *Annals and Antiquities of Rajasthan of the Central and Western States of India.* London: Oxford University Press, 1920.
27. "Umiam Hydel Project." *Indian Railway Gazette* 62, no. 9 (September 1964), 239.
28. United States Department of Commerce. *The Electrical Equipment Market in India.* Washington, DC: Government Printing Office, 1927.
29. "Vast Marble Dams, Hidden for Centuries in Hills of India, Are Unsung Triumphs of Engineering." *Popular Science Monthly*, April 1922, 37–38.
30. Venkatarama, V. *Power Development in India: The Financial Aspects.* New Delhi: Wiley Eastern, 1972.

CHAPTER 8

Tarapur and the Atomic Age, 1959–1964

In the summer months of 1965, the monsoon lashed the coastline of Maharashtra north of Bombay—as it did to a greater or lesser degree every year. The storms fell with equal force on fishing villages, coastal salt works, coconut plantations, and the construction site of India's first nuclear power station, Tarapur Atomic Power Project.

By the beginning of September, the monsoon had receded, giving the personnel building Tarapur a chance to increase the pace of work. The effects of the monsoon had been planned into the construction schedule by General Electric, the primary contractor, and Bechtel, the main construction subcontractor. Everything was on schedule for completion of the project in three years.[1]

Then—WAR. In the first days of September 1965, planes from the Indian and Pakistani air forces began clashing along the border between the two countries. On September 6, the conflict turned into open war. As Indian and Pakistani planes raided cities and airfields in each other's territory, local governments on both sides of the border responded by

[1] GE, TAPS monthly report 15, August 1965, box 21, USAID Closed Project Loan Files 1960–1971, RG 286, NARA.

W. A. T. Logan, *A Technological History of Cold-War India, 1947–1969*, Palgrave Studies in the History of Science and Technology,
https://doi.org/10.1007/978-3-030-78767-7_8

enacting blackout regulations. In Maharashtra, the cities of Bombay and Poona ordered that cinema halls close at 6:00 every evening.[2]

Sixty miles north of Bombay, Tarapur Atomic Power Project (TAPP) was well within the range of American-made Pakistan Air Force B-57 Canberra bombers. Fearing attack on this strategic project, TAPP authorities instituted a blackout at the project site. As there were no blackout shades in project housing, residents of the housing colonies could do little more than sit in darkness in the evening until it was time to go to bed. Starting on September 7, the TAPP authorities also shortened the second shift at the construction site, eliminating all work during night hours. From September 16 until the ceasefire on September 22, they canceled the second shift entirely.[3]

Pakistani planes never attacked Tarapur, but the conflict nevertheless did affect the project directly. At the outbreak of fighting, two American cargo ships carrying parts for the Tarapur project happened to be docked at Karachi harbor: SS *Express* and SS *Steel Vendor*. When the fighting began, Pakistani authorities boarded the ships and ordered that their cargoes be unloaded. *Steel Vendor* was carrying simple parts that could easily be replaced, like steel plates for lining the suppression pool. But the cargo hold of *Express* contained twenty-five tons of materials for the drywells, including a specialized hatch known as the personnel lock. Over protests by the ships' crews, the Pakistani authorities confiscated the parts destined for Tarapur before sending the ships on their way.[4]

Requests through diplomatic channels failed to get the parts returned, and the project authorities ended up placing new orders from the United States. What the Pakistanis did with the parts they kept is not clear. "So far as anybody knows," Rep. Craig Hosmer of California claimed in the US House of Representatives the following year, the reactor parts "are being used as hitching posts for camels." (This was, of course, pure speculation, and it says more about the congressman than the Pakistanis. It is possible that the parts ended up being used in Pakistan's own nuclear program.)[5]

[2] "Maharashtra: Magnificent Response," *Link*, September 19, 1965, 24.

[3] GE, TAPS monthly report 16, September 1965, box 21, USAID Closed Project Loan Files 1960–1971, RG 286, NARA; TAPP quarterly report 4, July–September 1965, box 20, ibid.; Arun Joshi, interview by author, Kennewick, WA, October 5, 2017.

[4] GE, TAPS monthly report 16; TAPP quarterly report 4.

[5] Cong. Rec., 89th Cong., 2nd sess., 1966, vol. 112, 11216–17.

The confiscation of parts for Tarapur demonstrates the complicated interplay between technology and international relations in South Asia during the Cold War. Pakistan, still a nominal American ally, had confiscated goods from an American flag-carrying ship, apparently for no other reason than to spite India. Time and again, projects of development and technological indigenization would get snared in the tangled web of regional and global relations of the Cold-War world.

More than any of the other topics studied in this book, Tarapur existed in the Cold-War world. Although TAPP was just one of several Indian power projects financed by the United States in the early 1960s (Umiam Hydroelectric Project was another), it received focused attention in national and international news. The governments of both India and the United States gave priority to parts and personnel needed for the project. The US government considered Tarapur a prestige project, the flagship of benevolent American efforts to help India and other former colonies transform themselves into modern nations with (capitalist) industrial economies. The Indian government also considered Tarapur to be a prestige project, as by adopting nuclear power—one of the highest-of-high western technologies—India would be demonstrating to the world that it had become modern at last.[6]

At the same time, Tarapur was an Indian project in India. Although implemented as a turn-key project by General Electric and Bechtel, the Indian government directed the project from beginning to end, from site selection to choosing the prime contractor all the way through the completion of construction. Tarapur was at once foreign—designed by engineers in California—and Indian—built by men and women carrying loads of rocks and cement on their heads under the hot tropical sun. Because of this dichotomy, Tarapur divides neatly into two chapters. In this chapter, we will consider Tarapur's place in the Cold-War world, and how an electrical engineering company from upstate New York and a construction company from San Francisco came to build a nuclear power station on a jungly stretch of Indian coast north of Bombay. In the final chapter, we will shift our attention to the technological and labor aspects of the project, to see what the lived experience was like for the multinational workforce that built Tarapur.

[6] An earlier Indian prestige project was Bhakra Dam. Ellsworth Bunker referred to Bhakra as such in a letter to Adolph Ackerman, December 15, 1959, box 3, USAID Mission to India Subject Files 1955–1962, RG 286, NARA.

The Nuclear Age Comes to Independent India

The Indian nuclear program exploded onto the world stage on May 18, 1974 when the Indian government detonated its first atomic bomb under the Rajasthan desert. This test, codenamed darkly "Smiling Buddha," was commonly referred to in press reports simply as PNE—Peaceful Nuclear Explosion. The Indian government claimed that the test was meant only to investigate the potential uses of nuclear explosions for nonmilitary purposes such as earthmoving or mining.

Although deliberately misleading, this claim was not as preposterous as it might now seem in the twenty-first century. The Indian government was not the first to suggest that nuclear explosions could be used for peaceful purposes. Project Orion, a study initiated by General Atomics of California in 1958, explored the possibility of using controlled nuclear explosions for rocket propulsion. From the late 1950s to the early 1970s, the US Atomic Energy Commission proposed employing nuclear explosions in large civil engineering projects such as forming an artificial harbor in Alaska or excavating a section of the Tennessee-Tombigbee Canal in Mississippi. Although these schemes were never executed, the USAEC did perform a series of nuclear tests under a project codenamed Plowshare, in reference to Isaiah 2:4: "They shall beat their swords into plowshares, and their spears into pruning hooks."[7] Despite these precedents, though, domestic and international observers correctly perceived that the Indian test was part of a weapons program designed to challenge Pakistan.

Ever since 1974, scholars in India and the West have been fascinated by the fact that India was able to develop one of the twentieth century's highest-of-high technologies less than thirty years after independence. Fascination has turned into preoccupation, so that nuclear technology is probably the aspect of postindependence Indian technology that has received the most scholarly attention, most of which has been focused on the Indian atomic bomb project rather than comparatively more mundane atomic research or nuclear power. In context, the Indian atomic bomb is hardly a miracle. Even though the Indian state had long disavowed any

[7] See Scott Kirsch, *Proving Grounds: Project Plowshare and the Unrealized Dream of Nuclear Earthmoving* (New Brunswick, NJ: Rutgers University Press, 2005); George Dyson, *Project Orion: The True Story of the Atomic Spaceship* (New York: Henry Holt, 2003).

intention of building nuclear weapons, it had had an atomic research organization since the year after independence. Furthermore, as Itty Abraham notes, no nation's nuclear program was entirely domestic; every program relied on technical input from other countries. During the Cold War, Soviet spies stole nuclear secrets from the Americans, the Soviets in turn contributed to the Chinese program, and scientists from all over the world received training in nuclear physics at American universities. As a recipient of technical aid for reactor construction from Britain, Canada, France, and the United States, India was no exception.[8]

Just two years before India gained independence, the world's most fearsome weapon made its debut. Developed in secret by the US government in a crash-program codenamed the Manhattan Project, the atomic bomb used nuclear fission to create an explosion equivalent to thousands of conventional bombs. In August 1945, two US Army Air Force B-29 bombers dropped atomic bombs on the Japanese cities of Hiroshima and Nagasaki. In Hiroshima, forty thousand people were killed in an instant when the bomb exploded above the city. Within days, another hundred thousand had succumbed to burns or radiation poisoning, and by the end of the decade the number of deaths had risen to 200,000—more than half the total population of the city. Of the 76,000 buildings in the city, 48,000 were completely destroyed. Property losses caused by the one bomb were equivalent to 1.1 million people's annual incomes. In the words of a Japanese study of the destruction wrought by the bombs, "The whole of society was laid waste to its very foundations."[9]

This technology would define international politics for decades. With it, humanity had gained for the first time in history the ability to commit planetwide collective self-immolation. Yet if properly contained, nuclear technology held great potential as an energy source. After the war, the major technological powers scrambled to exploit nuclear energy for peaceful as well as military purposes. Following initial small-scale experiments by both superpowers, the first plant to generate power for commercial use came online in the Soviet Union in 1954. Three years later, the United States commissioned its first commercial nuclear plant, the first devoted solely to producing electric power, as the station in

[8] Itty Abraham, *The Making of the Indian Atomic Bomb: Science, Secrecy, and the Postcolonial State* (London: Zed Books, 1998), 9–10.

[9] Richard Rhodes, *The Making of the Atomic Bomb* (New York: Simon & Schuster, 1986), 728–34.

Russia had also been made to produce plutonium. Great Britain, France, and Canada also began nuclear power programs in the early years after the war.[10]

Not wanting to be left behind in the race to exploit this new technology, the Government of India founded the Atomic Energy Commission in 1948. Six years later, the government created the Department of Atomic Energy (DAE) as a branch of the Atomic Energy Commission. The same year, DAE established its first dedicated atomic research laboratory, Atomic Energy Establishment, Trombay (AEET) in Bombay, under the leadership of Dr. Homi Jehangir Bhabha.[11]

A Cambridge-educated physicist, Bhabha had initially tried doing research in Europe. Although he was rich, cultured, and charismatic—as well as a talented theoretical physicist—he found that he could not break the colonial glass ceiling in Europe. Rather than whiling away his career as an underpaid and under-appreciated assistant in someone else's lab, he elected to return to his native India where race would not be an issue and he could get full support for his work. In India, Bhabha became a celebrity, one of the leading spokesmen of India's modernization. Much eulogized after his death in a plane crash in 1966, his picture or bust is a common sight at science museums around India to this day.[12]

Shortly after its establishment, AEET built Apsara, India's first nuclear reactor. First going critical in 1956, Apsara was a small 1-MW swimming pool-type reactor, in which the entire reactor assembly was submerged in a tank of water. Water is an especially good shield against radiation, which is also why spent nuclear fuel is stored in holding tanks. (According to the blog *What If?* by Randall Munroe, there is actually less radiation in the upper levels of a spent fuel pool than outside, because the water shields against radiation from the sun as well as from the bottom of the pool.) The open design of a swimming pool-type reactor also allows for easy access and servicing. Too small to be used for commercial power

[10] For nuclear power in the Soviet Union, see Paul R. Josephson, *Red Atom: Russia's Nuclear Power Program from Stalin to Today* (Pittsburgh: University of Pittsburgh Press, 2000); for France, see Gabrielle Hecht, *The Radiance of France: Nuclear Power and National Identity after World War II* (Cambridge, MA: MIT Press, 2009).

[11] Department of Atomic Energy, *Annual Report 1954–55* (n.p.: Government of India, 1955), 1–2. AEET was renamed Bhabha Atomic Research Centre after Homi Bhabha's death.

[12] Itty Abraham deconstructs Bhabha's legacy in *Making of the Indian Atomic Bomb*, 36–43.

generation, Apsara served as a research reactor for fifty-four years, until its final shutdown in 2010.[13]

In 1955, Canada signed an agreement to provide India a 40-MW reactor under the Colombo Plan. The reactor, dubbed Canada–India Reactor or CIRUS, used natural uranium as fuel and heavy water (D_2O or deuterium oxide) as a neutron moderator. The first loading of fuel also came from Canada, and the United States supplied the heavy water. Although a degree of magnitude larger than Apsara, CIRUS was still too small to be used for much besides research. The reactor produced plutonium as a byproduct, and under the Indo-Canadian agreement, the Indian government did not have to account for how this plutonium was used. Perhaps unsurprisingly, the PNE would end up using some of this plutonium.[14]

In these heady Nehruvian days, it was virtually a given that the DAE would be unsatisfied with importing nuclear reactors and fuels, and the nuclear program would eventually have to be indigenized. Early steps in this direction were taken by the DAE's Raw Materials Division, later the Atomic Minerals Division, which had the job of developing domestic sources for such useful minerals as uranium, zircon, and lithium. Accordingly, the division sent prospectors from one end of India to another, hunting for signs of radioactivity in the earth. Government prospectors descended into abandoned copper mines in Rajasthan, drove jeeps outfitted with radiation detectors in the mountains of Jammu & Kashmir, and performed aerial surveys of thousands of square miles with aircraft provided by the IAF. Private citizens also sent in soil and rock samples, hoping to receive prizes for discovering new deposits of radioactive minerals.[15]

The most promising deposits of uranium ore turned up in the Singhbhum area of Bihar (later Jharkhand). With assistance from two Soviet mining engineers, the DAE set up Jaduguda Mines, which started

[13] "Reactors," *Bhabha Atomic Research Centre*, http://barc.gov.in/reactor/index.html (accessed March 7, 2014); Randall Munroe, "Spent Fuel Pool," *What If?*, January 22, 2013, https://what-if.xkcd.com/29/ (accessed October 14, 2019).

[14] Robert Gillette, "India: Into the Nuclear Club on Canada's Shoulders," *Science*, June 7, 1974, 1053–55.

[15] Department of Atomic Energy, *Annual Report 1954–55* (n.p.: Government of India, 1955), 10–12; Department of Atomic Energy, *A Brief Report of the Activities of the Department of Atomic Energy for the Year 1958–59* (n.p.: Government of India, 1959), 16–17.

exploiting the main lode of the Singhbhum belt in 1967. The DAE also built an on-site uranium mill to process the ore into yellowcake, a powder that could, with further processing at another facility, be turned into UO_2 pellets for use in a power reactor.[16]

The Genesis of Tarapur

By the end of the 1950s, nuclear power technology had begun to circulate across the industrialized world. At the turn of the 1960s, the Soviet Union, Britain, the United States, and France were already operating nuclear plants, while Belgium, Canada, West Germany, Italy, and Sweden had plants under construction. In the first year of the new decade, nuclear power technology first arrived in Asia, when construction of the Japan Power Demonstration Reactor began.

At about the same time, in August 1959, the Indian Department of Atomic Energy established its Power Unit, making India the first nation outside the highly industrialized world to initiate a nuclear power program. Fifty-eight-year-old M.N. Chakravarti, who had spent decades working for the railways—most recently as General Manager of the Western Railway—was tapped to lead the unit. Precocious engineer M.R. Srinivasan, who held degrees from Mysore Engineering College and McGill University in Montreal and was still in his twenties, was picked to serve as Principal Project Engineer. Maheshwar Dayal joined the team as Project Engineer (Electrical). Constructing a nuclear power station became an official goal of the Planning Commission when it issued the Third Five-Year Plan in 1961. The Plan called for the construction of a plant at Tarapur with two reactors, each producing 150 MW of power. Rs. 510 million would be allocated to the project.[17]

Planners in the DAE had several rationales for initiating an atomic power project. Official publications explained the decision as the outcome of a simple cost–benefit analysis: industrialization along the Indian west

[16] Department of Atomic Energy, *Brief Report for 1961–62* (Bombay: Government Central Press, 1962), 8–10; Department of Atomic Energy, *Brief Annual Report 1962–63* (n.p.: Government of India, 1963), 13.

[17] "Central Railway General Manager: Mr. Chakravarti Takes Charge," *TOI*, January 18, 1955; Planning Commission, "Third Five Year Plan," http://planningcommission.nic.in/plans/planrel/fiveyr/3rd/welcome.html (accessed January 13, 2014); M.R. Srinivasan, "An Intermezzo," in *Selected Lectures of Dr. M.R. Srinivasan* (Bombay: Department of Atomic Energy for the Nuclear Power Corporation of India, n.d. [1990]), 1–6.

coast was creating ever-increasing demands for electricity, and nuclear was the only power source that made economic sense. According to the official explanation, since India's coal resources were mainly on the eastern side of the country, transporting this coal to the west coast would be prohibitively expensive. In addition, the region's hydropower potential was already mostly tapped-out. Nuclear was thus the only option left.[18]

The same official publications implied, but never formally acknowledged, an additional motive: the planners wanted their country to have nuclear power because they saw it as higher-tech and more modern, and therefore more desirable, than any of the other options. An anonymous writer, reflecting on the dedication of Tarapur Atomic Power Station in 1970 for *The Indian and Eastern Engineer*, expressed such unbridled technological enthusiasm by writing: "As we looked around the quiet surroundings [of Tarapur Atomic Power Station], the serene buildings, the sea to the west and lush green lawns below the 112m [370-ft] ventilation stack, as we looked at the august gathering and mingled with them freely, we could sense the feeling of partnership, a feeling of triumph at the success of another venture in the path of human progress."[19]

The DAE set out to find a site for its nuclear power project. A key consideration was providing a means to moderate the temperature of the reactor. Nuclear plants often have huge hyperboloid cooling towers for this reason. In pop culture, cooling towers can be synonymous with nuclear plants (think of Mr. Burns's plant in *The Simpsons*), but the Power Unit ruled out their use because they are not efficient in hot climates. Instead, the plant would have to be built at a site with access to large quantities of water. This meant a coastal site, because rivers in western India are seasonally unreliable. After studying maps of 200 miles of coastline between Bombay and the Gulf of Cambay in Gujarat, the Power Unit decided to investigate four sites. Shortly after the formation of the Power Unit in 1959, M.N. Chakravarti and M.R. Srinivasan set out to scout the sites. Survey teams followed behind them and performed hydrographic surveys and exploratory drilling at each location. Tarapur Point, where the North Konkan Coast juts out into the Arabian Sea north of Bombay,

[18] M.N. Chakravarti and M.R. Srinivasan, "Selection of Site for the Tarapur Atomic-Power Station." In International Atomic Energy Agency, *Siting of Reactors and Nuclear Research Centres* (Vienna: International Atomic Energy Agency, 1963), 458.

[19] "Dedication of Tarapur Nuclear Power Plant," *The Indian and Eastern Engineer*, February 1970, 113.

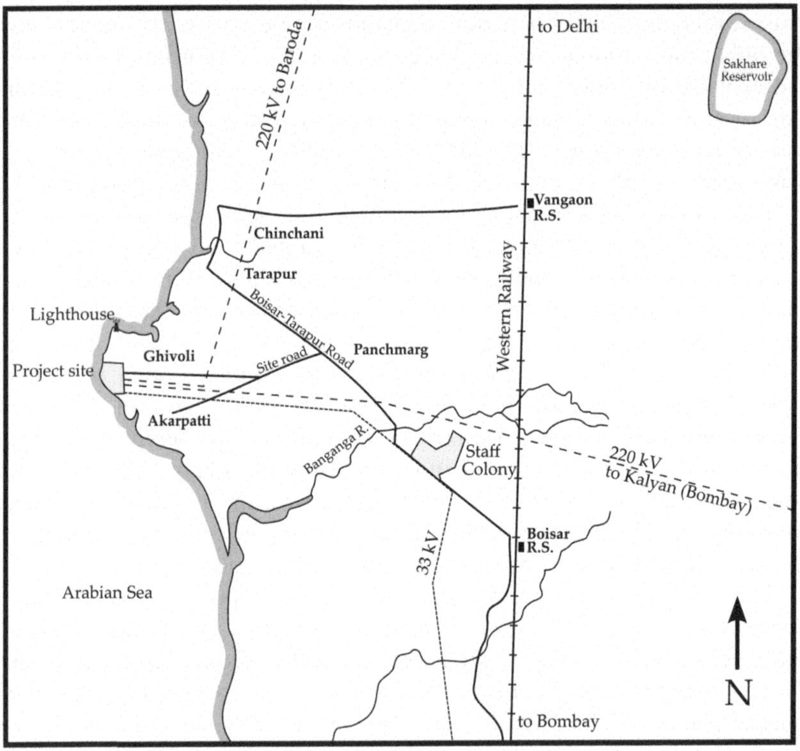

Fig. 8.1 Map of the North Konkan Coast around Tarapur (Map drawn by the author, based on map in M.N. Chakravarti and M.R. Srinivasan, "Selection of Site for the Tarapur Atomic-Power Station," in International Atomic Energy Agency, *Siting of Reactors and Nuclear Research Centres* [Vienna: International Atomic Energy Agency, 1963], 457–65)

emerged as the most favorable site (Fig. 8.1). Since it was located on the Deccan Trap landform, the location was not susceptible to earthquakes. Bedrock was close to the surface, with sound basalt appearing in the drillings as close as fifteen feet below ground level. Crucially, the site was close to a Western Railway railhead at Boisar.[20]

[20] Chakravarti and Srinivasan, "Selection of Site for the Tarapur Atomic-Power Station," 457–65.

Three decades later, when M.R. Srinivasan recalled his first visit to the site, he portrayed it as an isolated, desolate, monsoon-swept stretch of coast, accessible only by bullock-cart.[21] In fact, the area was well-connected by rail (as the Power Unit knew), and it already had a long history of human habitation. In 1534, the Sultan of Gujarat ceded the North Konkan (the section of coast including Tarapur) to the Portuguese. In order to control the coast and defend against rival powers that they considered pirates, the Portuguese built defensive garrisons at various strategic positions along the coast, including Tarapur. In 1593, under the viceroyalty of Matias de Albuquerque, the defensive post at Tarapur was upgraded to a proper fort, with stone walls thirty feet high and ten feet thick. The Portuguese held their own against the Mughals in the North Konkan, but they were unable to resist the next great power of India, the Maratha Empire. In 1739, Maratha forces under the command of Chimnaji Appa seized the the North Konkan from the Portuguese. The Marathas repaired Tarapur Fort, adding some Indian architectural flourishes to an otherwise European structure.[22]

With the signing of the Treaty of Poona in June 1817, the North Konkan passed into the hands of the British after nearly eighty years under Maratha rule. Until Indian independence, the Tarapur area was directly administered by the British as part of the Bombay Presidency. With independence in 1947, the presidency became Bombay State. In 1960, after riots in Bombay, Ahmedabad, and other cities, Bombay State split on linguistic lines: Gujarati-speaking Gujarat in the northwest and Marathi-speaking Maharashtra in the southeast. After the reorganization,

[21] Srinivasan, "An Intermezzo," 6. Itty Abraham criticizes Srinivasan's account for representing the "rational, scientific modern emerging from the *tabula rasa* (or rather, the slate *made* blank) of the traditional Indian landscape." Abraham, *Making of the Indian Atomic Bomb*, 161.

[22] Alexander Kyd Nairne, "History of the Konkan," in *Gazetteer of the Bombay Presidency: History of the Konkan Dakhan and Southern Maratha Country* (Bombay: Government Central Press, 1896), 40, 46, 54, 84, 97, 115, 119; C.K. Srinivasan, *Baji Rao I: The Great Peshwa* (Bombay: Asia Publishing House, 1961), 112–32. An inscription on the northern wall of Tarapur Fort declares, "This fort was ordered to be built by the illustrious Sir Matias de Albuquerque, viceroy of India, in the year 1593." (Translation from the Portuguese by Lucas Machado.)

Tarapur village fell within the new state of Maharashtra, but it was only twenty-five miles south of the Gujarat border.[23]

When Chakravarti and Srinivasan scouted the Tarapur site shortly before the reorganization of Bombay State, the area was home to thousands of people. Most of them lived in the twin villages of Tarapur and Chinchani, which straddled a creek northwest of the chosen site for the power station on Tarapur Point. According to the 1961 census, the combined population of these two villages was over 12,600. The main occupations of the residents of Tarapur, Chinchani, and other nearby villages were fishing, agriculture, and poultry-rearing. Tarapur also had a sea-salt works, and the residents of Uchali and Kelwa farmed oysters. Although the Tarapur area was not modern or industrialized before the construction of the nuclear power station, it was an economically vibrant area, connected by trade to the rest of India.[24]

The Race for the Tarapur Contract

On October 10, 1960, M.N. Chakravarti formally announced an international bidding competition for the selection of a prime contractor to build a 300-MW nuclear plant running on natural uranium for the Department of Atomic Energy. Representatives of interested firms were instructed to visit Chakravarti's office on Apollo Pier Road in Bombay to obtain a copy of the four-part enquiry specifications, for a nonrefundable fee of Rs. 1,000 (US$210 at the 1960 exchange rate). Although the specifications called for natural-uranium reactors, the tender notice stated that firms offering enriched-uranium reactors could, at their own risk, bid for the contract, provided that they clearly specified that their bid was for an enriched-uranium plant.[25]

One of the companies that expressed interest in the Tarapur tender notice was the American firm General Electric. The requirement for

[23] Nairne, "History of the Konkan," 115; Bipan Chandra, Mridula Mukherjee, and Aditya Mukherjee, *India Since Independence* (New Delhi: Penguin Books, 2008), 128–29.

[24] P.R. Kamath, I.S. Bhat, A.A. Khan, and A.K. Ganguly, "Preoperational Search for Baseline Radioactivity, Critical Food and Population Group at the Tarapur Atomic Power Station Site," paper presented at the First International Congress of IRPA (Rome, September 5–10, 1966), Government Publications, NLI.

[25] TAPP, Tender Notice, box 9, US Mission to IAEA Classified Records 1957–1963, RG 84, NARA.

natural uranium would be a sticking point, though, because GE, like its competitor Westinghouse, made enriched-uranium reactors.

In the 1960s, General Electric was actively marketing its brand of boiling water reactors (BWR) internationally. Bidding for Tarapur would be just a part of the expansion of the brand, as GE ultimately sold BWRs to ten countries, including Japan, Germany, and Mexico. In preparing to move into the Indian market, the company placed advertisements in periodicals such as the *Indian Journal of Power and River Valley Development*, highlighting its international experience. One such ad, titled "Excellent progress being made around the world on General Electric boiling water reactor projects," appeared in the October 1961 issue of the journal. The two-page spread featured a photograph of the 160-foot containment sphere of the Garigliano Nuclear Power Station in Italy, which was purportedly "twice the size of St. Peter's dome." The ad copy also mentioned the Tokai Mura plant in Japan and Big Rock Point and Humboldt Bay projects in the United States. "Long identified as a leader in atomic development," the copy boasted, "G.E. is engaged in a widely diversified program of nuclear application. BWR plants are now being offered up to 500 megawatts, incorporating such advanced concepts as high-power density cores, pressure suppression containment and internal steam separation."

On the international market, two different types of American reactors, GE's BWRs and Westinghouse's pressurized water reactors (PWR) were in direct competition with gas-graphite reactors from France and Britain, and CANDU pressurized heavy-water reactors from Canada. All the different types of reactors represented slightly different approaches to nuclear power generation, based on whether they used light or heavy water for the reactor moderator, and enriched or natural uranium as reactor fuel. Both BWRs and PWRs use enriched uranium, which has a higher portion of the isotope uranium-235 than natural uranium, which is more than 99 percent uranium-238. Enriched uranium is produced in hugely expensive gaseous diffusion plants. At the time, the only western-bloc facility capable of enriching uranium on a large scale was located at Oak Ridge, Tennessee. The Soviet Union also produced enriched uranium for its reactors, but the quarantining of eastern- and western-bloc technology in early independent India ensured that Soviet uranium would not be available for the Tarapur project.[26]

[26] M.N. Chakravarti and M.R. Srinivasan, "Selection of Site for the Tarapur Atomic-Power Station," in International Atomic Energy Agency, *Siting of Reactors and Nuclear*

In deciding whether to bid on Tarapur, GE's representatives felt at turns pessimistic and cautiously optimistic about their prospects of getting the contract. Robert King of GE's New York office and William Lalor, Jr. of the Atomic Power Equipment Department in San Jose traveled to India in January 1961 to gather information on the project. They spent several days talking with Chakravarti, and also visited the Tarapur site. The American Consul General in Bombay, Sidney Sober, reported in a foreign service dispatch that King and Lalor would not make a decision on whether their company should bid until Homi Bhabha guaranteed that their submission would be given a "fair chance." When the men met with Bhabha in New Delhi and asked whether natural-uranium reactors would be preferred over enriched-uranium reactors, Bhabha replied, "I think both types will receive equal consideration." Feeling reasonably assured, they tentatively decided to recommend to company headquarters that GE should prepare a bid. Within days, though, Bhabha had appeared before the Indian press and hinted strongly that he hoped India's first nuclear powerplant would use natural uranium. To assuage the fears of the American companies, Chakravarti wrote a letter dated March 29, 1961 to both GE and Westinghouse. The letter made no mention of Bhabha's assurance of "equal consideration," but it was enough to convince GE and Westinghouse to submit bids.[27]

But as the bidding deadline of August 31, 1961 approached, prospects for the American companies, especially GE, again began to look bleak. Consul General Sober declared in a foreign service dispatch that "the Indian DAE clearly hopes to have the Tarapur station built by a British firm supplying natural uranium reactors." To make matters worse, the US government announced that it would not finance the project. But by this point, the two American companies had spent significant time and

Research Centres (Vienna: International Atomic Energy Agency, 1963), 458; M.N. Chakravarti, "Tarapur Atomic Power Project," *IJPRVD* 18, no. 2 (February 1968), 41.

27 "General Electric interest in proposed atomic power plant; related notes on India's nuclear development program," foreign service dispatch 248, Amcongen Bombay to Department of State, January 30, 1961, box 1, USAID Classified Central Subject Files 1953–1962, RG 469, NARA; "Proposed atomic power plant at Tarapur," foreign service dispatch 489, American Consul General Bombay to Department of State, February 24, 1961, box 9, US Mission to IAEA Classified Records 1957–1963, RG 84, NARA; "Recent developments concerning proposed atomic power plant at Tarapur," foreign service dispatch 581, American Consul General Bombay to Department of State, April 7, 1961, ibid.

resources preparing bids, and they felt that it was too late to quit. Representatives of the Indian subsidiaries of GE and Westinghouse delivered their companies' bids to the Bombay project office by the August 31 deadline.[28]

Chakravarti's office received a total of seven bids from companies representing four different western countries. (As usual for Indian development projects of this time, the Soviet Union declined to participate when western-bloc firms announced their intention to submit bids.) English Electric, General Electric Company of the United Kingdom, and an ad hoc French consortium submitted bids for natural uranium gas-graphite reactors. Canadian General Electric Company and Atomics International (an American firm) also proposed natural-uranium reactors, although the DAE rejected these tenders because they did not provide firm cost estimates. Westinghouse and General Electric submitted the only two bids for enriched-uranium reactors.[29]

After General Electric had experienced so much angst about whether it was even worth bidding on the project, it came as a complete surprise when the DAE issued a letter of intent to GE, announcing that its bid had emerged as the most favored of the seven received. GE had underbid the

[28] Foreign service dispatch 581; "American bids on Tarapur Nuclear Power Plant," air pouch, American Consul General Bombay, October 3, 1961, box 9, US Mission to IAEA Classified Records 1957–1963, RG 84, NARA. G.B. Sujan was the Westinghouse Regional Representative in Bombay who delivered his company's bid. According to Consul General Sober, Sujan reported to the consulate that "he had presented to the Indian authorities a plastic model of a single-reactor plant about 30″ square and 10″ high, scaled at 1/16″ to the foot [1/192]. This model, which is not necessarily a true model of a plant that Westinghouse might build at Tarapur, is said to have made a hit with the Indian officials."

[29] M.N. Chakravarti and M.R. Srinivasan, "Tarapur Atomic Power Station," in *Proceedings of the Third International Conference on the Peaceful Uses of Atomic Energy*, vol. 5: *Nuclear Reactors-I. Gas-Cooled and Water-Cooled Reactors* (New York: United Nations, 1965), 192–98; "Tarapur/Oyster Creek Comparison," n.d., Tarapur project summary file, box 24, USAID Closed Project Loan Files 1960–1971, RG 286, NARA. English Electric submitted its bid in collaboration with another British firm, Taylor Woodrow Construction, and the Indian firm Hindustan Construction Company. Although English Electric did not receive the contract, HCC later built the foundations for Rajasthan Atomic Power Station. Hindustan Construction Co., "Thirty-sixth Directors' Report and Statement of Accounts for the Year Ended 31st August 1961," S. No. 240, Hindustan Construction Company subject files, papers of Walchand Hirachand, NMML.

other companies, but still managed to offer 380 MW of net power—80 MW more than required by the enquiry specifications.[30]

Although the DAE's letter of intent was still subject to intergovernmental agreements regarding financing and the transfer of nuclear equipment and materials, this was a crucial first step. The US government soon reversed its earlier decision and agreed to finance the foreign-exchange costs of the project with an $80-million USAID loan authorized on June 28, 1963. The loan agreement had many of the same provisions as the agreement for Umiam Hydroelectric Project, including the small business notification clause.[31]

Representatives of the Indian and US governments signed the formal intergovernmental agreement for the project on August 8, 1963. The following May, the Government of India signed a contract with General Electric and its Indian subsidiary that called for the completion and handover of Tarapur Atomic Power Station within fifty-two months of the contract date.[32]

Almost, but not everything, was in place for the project to get underway. One final matter needed to be resolved: the question of fuel supply and safeguards. As a charged cold-war issue, this problem threatened to undo the Tarapur agreement entirely.

[30] M.N. Chakravarti, "Tarapur Atomic Power Project," *IJPRVD* 18, no. 2 (February 1968): 41–47.

[31] "Loan Agreement (India: Tarapur Nuclear Power) between the President of India and the United States of America," December 7, 1963, file no. 2(21) AID/69, Ministry of Finance, NAI; Tarapur Atomic Power Project contract between the President of India and General Electric for the construction of an atomic power station to be located near Tarapur, final draft, April 10, box 12, USAID Closed Project Loan Files 1960–1971, RG 286, NARA. In 1967, USAID deobligated $5 million of the loan amount. Implementation Letter No. 6, Daniel G. Pfoutz to Finance Secretary, January 13, 1969, file no. 2(21) AID/69, Ministry of Finance, NAI.

[32] Cong. Rec., 88th Cong., 1st sess., 1963, vol. 109, 15221–25; Tarapur Atomic Power Project contract between the President of India and General Electric for the construction of an atomic power station to be located near Tarapur, final draft, April 10, 1964, box 12, USAID Closed Project Loan Files 1960–1971, RG 286, NARA.

Tarapur and the Cold-War World

Why had General Electric been chosen as the prime contractor for Tarapur Atomic Power Station? If the Indian Department of Atomic Energy had expressed a preference for natural-uranium reactors, why was GE chosen to deliver reactors running on enriched uranium? Like so many other decisions in Indian development at this time, technical, economic, and ideological concerns pulled in different directions, and making this decision entailed balancing these opposing concerns. Ordering a reactor from GE was a compromise that offered India a reliable supply of nuclear fuel and access to diverse technical knowledge.

Whether it runs on enriched or natural uranium, a nuclear reactor needs a large amount of fuel to sustain a reaction and produce power. For example, the first fuel load of Tarapur, installed in 1969 and left in place for two and one-half years, consisted of forty metric tons (88,000 lbs) of enriched uranium per reactor. A reactor's fuel assembly consists of numerous fuel rods (284 per reactor in the case of Tarapur), which are in turn filled with small fuel pellets stacked on top of each other. The pellets consist of a uranium core clad with some other material (Tarapur's pellets used zircalloy, an Indian innovation), and they are small enough to be held between the thumb and forefinger of a (gloved) human hand—which is how they often are portrayed in science books and museum displays. The uranium in the pellets, in turn, is derived from yellowcake, which is extracted from uranium ore dug out of a mine.[33]

At the time of the bidding for Tarapur, the Indian government was starting to develop facilities for producing nuclear fuel domestically. The Jaduguda Mine and Mill were in planning stages, and they would begin production in 1967. A facility for producing uranium dioxide (UO_2) and fabricating fuel pellets was farther off, but work on the Nuclear Fuel Complex in Hyderabad would begin in 1968.[34] By 1961, India thus had

[33] M. Dayal, "Experience from Operating and Fuelling a Nuclear Power Plant in an Industrially Developing Country," in *Experience from Operating and Fuelling Nuclear Power Plants: Proceedings of a Symposium on Experience from Operating and Fuelling Nuclear Power Plants, Organized by the International Atomic Energy Agency and Held in Vienna, 8–12 October 1973* (Vienna: International Atomic Energy Agency, 1974), 85–87; M.N. Chakravarti, "Tarapur Atomic Power Project," *IJPRVD* 18, no. 2 (February 1968): 41–47; "Flexibility in the Design," n.d., Tarapur project summary file, box 24, USAID Closed Project Loan Files 1960–1971, RG 286, NARA.

[34] "History of NFC," *Nuclear Fuel Complex*, https://www.nfc.gov.in/history.html (accessed January 15, 2020).

the potential of producing natural-uranium fuel, but enriched uranium remained very much out of reach, because the reactors and other equipment needed for enriching uranium were a degree of magnitude more expensive than the facilities for natural uranium.

The DAE had initially planned to request bids only for natural-uranium reactors, but a clause permitting bids for enriched-uranium reactors was added to the tender notice on the request of Westinghouse. Even after this concession, members of the DAE disagreed over whether they should seriously consider enriched-uranium bids. In a discussion with US Consul General Sidney Sober in February 1961, H.N. Sethna, head of the Chemical Group at Atomic Energy Establishment Trombay, gave three reasons that made natural uranium a better fuel choice than enriched uranium. First, a reactor using natural uranium could produce plutonium, which would ultimately allow India to use domestically sourced thorium for power generation. Second, using natural uranium would avoid recurring foreign-exchange payments for imported fuel, once Jaduguda Mines and the Nuclear Fuel Complex were in operation. Third, a domestic source of fuel would be more reliable than a foreign source, which could be cut off during an emergency. In Sethna's view, these three reasons shifted the balance in favor of natural uranium. It is significant that all three reasons relate to autarky and self-reliance, calling on common themes of using indigenous materials, conserving foreign exchange, and protecting the security of strategic assets.[35]

Compared to Sethna, M.N. Chakravarti favored a more pragmatic approach. He had less of an ideological commitment to autarky and thought that the deciding factors for choosing between the bids should be overall cost and foreign-exchange component. Chakravarti's own quick calculations suggested that the up-front cost of an American enriched-uranium plant would be lower than a natural-uranium plant, with a foreign-exchange savings of $35 million. According to Consul General Sober, Chakravarti had told him that "it is his job to get the power station built and he is not interested in making excuses why it might not be done."[36]

[35] "American bids on Tarapur Nuclear Power Plant," October 3, 1961, Foreign service dispatch 489.

[36] Foreign service dispatch 489.

Chakravarti's pragmatic approach prevailed with the selection of the bid from General Electric. Although the Indian government was reticent to buy a plant that ran on imported enriched uranium, the US government guaranteed a long-term fuel supply over the course of twenty-five years, the planned lifespan of the plant. These terms proved convincing, and the Indian government agreed to buy American reactors that ran on American fuel.[37]

By selecting an American plant, India turned to a new donor nation and thus diversified its nuclear technology. Canada had become India's first donor of nuclear technology with the Canada–India Reactor, and India's second nuclear plant, then in early planning stages, would also be built with Canadian assistance. By overseeing the construction of an American plant in India, the specialists of the DAE would be exposed to a new set of tools and techniques, which they could apply toward their ultimate goal of producing an indigenous reactor.[38]

By selecting an American bid for the first plant, India was also diversifying risk. A downturn in Indo-Canadian relations could lead to the freezing of Canadian assistance, just as the United States had cut off military aid to India during the 1965 war with Pakistan. The United States might not cancel aid for the same reasons as Canada, so the risk of losing all support was lower if the aid came from multiple nations. As an editorial in the *Indian Journal of Power and River Valley Development* argued, "If the huge capital investment required for such stations are not immediately borne by India, it may be better if we make a start with two stations with the aid of two countries instead of with only one."[39]

This would prove to have been a wise strategy. After the Smiling Buddha atomic bomb test in 1974, Canada cut off all nuclear aid to India, leaving the second reactor at Rajasthan Atomic Power Station unfinished.

[37] M. Dayal, "Experience from Operating and Fuelling a Nuclear Power Plant in an Industrially Developing Country," 85–87; M.N. Chakravarti, "Tarapur Atomic Power Project," *IJPRVD* 18, no. 2 (February 1968): 41–47.

[38] Raja Ramanna, "Indian Nuclear Programme: Achievements and Prospects," *Journal of the Korean Nuclear Society* 19, no. 3 (September 1987): 213–19. My thanks to Arne Kaijser for being the first to suggest to me that ordering plants from two countries would give India a broader range of technical experience.

[39] "Policy for Atomic Power," editorial, *IJPRVD* 10, no. 2 (March 1960): 31.

On the other hand, the United States, while uneasy about the implications of the test, did not cut off the fuel supply to Tarapur.[40]

For the American government, the problem of fuel supply was inextricably linked to another issue: safeguards. One of the United States' greatest fears in the early Cold War was nuclear proliferation, which meant a non-nuclear state's gaining access to nuclear weapons, either by indigenous development or import. (The Indians were also concerned about nuclear proliferation, but they understood the term to mean current nuclear states such as the United States and Soviet Union expanding their nuclear weapons arsenals.) In 1945, the United States had had a global monopoly on nuclear weapons, but within a few years, the Soviet Union had joined the nuclear club. Western-bloc allies Britain and France would also join the club within the first decade and a half after the end of World War II.

The first four members of the nuclear club were part of the Cold-War balance of power, and after 1960 America's Cold-War leaders especially wanted to prevent nations outside the eastern and western blocs from gaining access to nuclear weapons. The United States was thus wary about supplying nuclear fuel to a country like India, because this could be a key step toward that country's developing nuclear weapons. Spent nuclear fuel could be reprocessed into plutonium, which in turn could be used in nuclear warheads.

Since the early 1950s, when it had become clear that the United States had lost its monopoly on nuclear weaponry, the Americans had been pushing for an international system of safeguards for atomic weapons and power facilities. In a 1954 speech before the General Assembly of the United Nations, President Eisenhower called upon the nuclear-capable countries to develop nuclear technology for the benefit of all people. "It is not enough to take this weapon [nuclear weaponry] out of the hands of the soldiers," he declared. "It must be put into the hands of those who will know how to strip its military casing and adapt it to the arts of peace." This speech, known as "Atoms for Peace," led to the creation of the United Nations International Atomic Energy Agency (IAEA), which was tasked with inspecting nuclear facilities internationally.[41]

[40] Robert Gillette, "India: Into the Nuclear Club on Canada's Shoulders," *Science*, June 7, 1974, 1053–55.

[41] Dwight D. Eisenhower, "Atoms for Peace," *Eisenhower Presidential Library*, http://www.eisenhower.archives.gov/all_about_ike/speeches/atoms_for_peace.pdf (accessed January 11, 2016).

Safeguards were a major source of contention in the negotiations between the United States and India before the construction of Tarapur. Homi Bhabha had desperately wanted to build Tarapur without any safeguards, and before the open call for tenders, he had tried to negotiate directly with the superpowers as well as western European countries to play them off of each other. But both superpowers did not want Tarapur built without safeguards, so they worked together to make sure that did not happen. The open call for tenders, in place of a bilateral agreement, was actually a defeat for Bhabha and an acknowledgment that safeguards were inevitable. What form these would take, though, remained open to negotiation.[42]

The United States wanted India to submit Tarapur to IAEA safeguards, to ensure that no fissile materials from the plant were ever diverted to a hypothetical nuclear weapons program, but India did not want to submit to IAEA inspections. As the *Times of India* explained in January 1963, IAEA inspections were "not acceptable to this country...since such international inspection is discriminatory in effect, with advanced nuclear nations being exempt from it." If reactors in France and Britain did not need to be subjected to safeguards, why should India be treated differently?[43]

Instead of international safeguards, India wanted to enter into a bilateral inspection agreement with the United States. After months of negotiations, the two countries reached a compromise and signed an agreement on August 8, 1963. The agreement stipulated that the United States would inspect the Tarapur plant at first, with a vague provision that inspections would shift to the IAEA someday. Under continued American pressure, India finally relented in 1971 and agreed to submit to IAEA inspections.[44]

[42] David C. Engerman, *The Price of Aid: The Economic Cold War in India* (Cambridge, MA: Harvard University Press, 2018), 149–56.

[43] "Current Topics: Safeguards," *TOI*, January 16, 1963.

[44] Cong. Rec., 88th Cong., 1st sess., 1963, vol. 109, 15221–25; Ajit Bhattacharjea, "Agreement on U.S. aid for Tarapur Reactor," *Hindustan Times*, June 27, 1963; "India Agrees to Inspection of Atomic Plant," *New York Times*, June 26, 1963; Gursharan S. Dhanjal, *Tarapur: The Politics of Nuclear Age* (Delhi: Rajdhani Book Service, 1992), 15.

Understanding Tarapur

From the beginning, Tarapur Atomic Power Station was contested and controversial, with its domestic and international stakeholders thinking about it in fundamentally different ways. The United States saw Tarapur as an opportunity to create closer economic and diplomatic ties with India. Although Ambassador John Kenneth Galbraith deemed the Indian government's plans for a nuclear plant near Bombay to be economically infeasible, he nevertheless thought that the United States should finance the project if given the chance. As he explained in a foreign service dispatch dated May 3, 1961, Tarapur "would be known as the 'American-aided' plant ... It would appeal to the educated groups, to the scientists, to the most progressive Indians, and to those to whom India's prestige in Asia is of primary importance. U.S. financing of this project would link the name of the United States with scientific achievement and advanced techniques."[45]

To the Indian bureaucrats and technocrats concerned with planning, financing, building, and operating the station, an American-designed plant would be little more than a stopgap, an intermediate step between colonial dependence and full indigenization of nuclear power technology. This was the ideal. The reality of course was that the construction of an enriched-uranium nuclear power station in India would establish a long-term dependency relationship with the United States. But broader diplomatic ramifications of this arrangement seem not to have been a major concern on the Indian side.

Tarapur is perhaps one of the great success stories of Nehruvian development and technology transfer. The Indian government got almost everything it wanted out of the project: a functioning power station, a guaranteed long-term supply of fuel, and knowledge transfer in the form of training for Indian builders and operators of the plant (discussed in the next chapter). But the project also represents the limits of nonaligned development in the Cold War. Because nuclear power bore kinship to nuclear weaponry, the export of nuclear technology and fuel to India was tightly controlled, and the Indian authorities could not operate the station or dispose of spent fuel entirely as they pleased.

[45] "Should the U.S. finance the nuclear power plant at Tarapur?" foreign service dispatch 1144, American Embassy New Delhi to Department of State, May 3, 1961, box 9, US Mission to IAEA Classified Records 1957–1963, RG 84, NARA.

By mid-1964, all the international agreements had been negotiated and signed. It was time for the construction of Tarapur Atomic Power Station to begin. Laborers, engineers, and managers, as well as parts, tools, and materials, had begun to converge on Tarapur Point for this great undertaking. When excavation began at the project site on September 13, 1964, no one knew that five years would elapse—during which one war and multiple labor strikes would take place—before the nuclear plant would finally be ready for regular operation.

References

1. Abraham, Itty. *The Making of the Indian Atomic Bomb: Science, Secrecy, and the Postcolonial State*. London: Zed Books, 1998.
2. Chakravarti, M.N. "Tarapur Atomic Power Project." *IJPRVD* 18, no. 2 (February 1968): 41–47.
3. Chakravarti, M.N., and M.R. Srinivasan. "Selection of Site for the Tarapur Atomic-Power Station." In International Atomic Energy Agency, *Siting of Reactors and Nuclear Research Centres*, 457–65. Vienna: International Atomic Energy Agency, 1963.
4. Chakravarti, M.N., and M.R. Srinivasan. "Tarapur Atomic Power Station." In *Proceedings of the Third International Conference on the Peaceful Uses of Atomic Energy, vol. 5: Nuclear Reactors-I. Gas-Cooled and Water-Cooled Reactors*, 192–98. New York: United Nations, 1965.
5. Chandra, Bipan, Mridula Mukherjee, and Aditya Mukherjee. *India Since Independence*. New Delhi: Penguin Books, 2008.
6. Dayal, M. "Experience from Operating and Fuelling a Nuclear Power Plant in an Industrially Developing Country." In *Experience from Operating and Fuelling Nuclear Power Plants: Proceedings of a Symposium on Experience from Operating and Fuelling Nuclear Power Plants, Organized by the International Atomic Energy Agency and Held in Vienna, 8–12 October 1973*, 85–103. Vienna: International Atomic Energy Agency, 1974.
7. "Dedication of Tarapur Nuclear Power Plant." *The Indian and Eastern Engineer*, February 1970, 113–15.
8. Department of Atomic Energy. *A Brief Report of the Activities of the Department of Atomic Energy for the Year 1958–59*. N.p.: Government of India, 1959.
9. Department of Atomic Energy. *Annual Report 1954–55*. N.p.: Government of India, 1955.
10. Department of Atomic Energy. *Brief Annual Report 1962–63*. N.p.: Government of India, 1963.

11. Department of Atomic Energy. *Brief Report for 1961–62*. Bombay: Government Central Press, 1962.
12. Dhanjal, Gursharan S. *Tarapur: The Politics of Nuclear Age*. New Delhi: Rajdhani, 1992.
13. Engerman, David C. *The Price of Aid: The Economic Cold War in India*. Cambridge, MA: Harvard University Press, 2018.
14. Gillette, Robert. "India: Into the Nuclear Club on Canada's Shoulders." *Science* (June 7, 1974): 1053–55.
15. Hecht, Gabrielle. *The Radiance of France: Nuclear Power and National Identity after World War II*. Cambridge, MA: MIT Press, 1998.
16. Josephson, Paul R. *Red Atom: Russia's Nuclear Power Program from Stalin to Today*. Pittsburgh: University of Pittsburgh Press, 2000.
17. Nairne, Alexander Kyd. "History of the Konkan." In *Gazetteer of the Bombay Presidency: History of the Konkan Dakhan and Southern Maratha Country*, 1–131. Bombay: Government Central Press, 1896.
18. Planning Commission. *Third Five Year Plan*. http://planningcommission.nic.in/plans/planrel/fiveyr/3rd/welcome.html (accessed January 13, 2014).
19. Ramanna, Raja. "Indian Nuclear Programme: Achievements and Prospects." *Journal of the Korean Nuclear Society* 19, no. 3 (September 1987): 213–19.
20. Rhodes, Richard. *The Making of the Atomic Bomb*. New York: Simon & Schuster, 1986.
21. Srinivasan, C.K. *Baji Rao I: The Great Peshwa*. Bombay: Asia Publishing House, 1961.
22. Srinivasan, M.R. "An Intermezzo." In *Selected Lectures of Dr. M.R. Srinivasan*, 1–11. Bombay: Department of Atomic Energy for the Nuclear Power Corporation of India, n.d. [1990].

CHAPTER 9

Constructing Tarapur, 1964–1969

On the evening of Wednesday, December 29, 1965, police rifles rang out in the darkness of the Konkan coast sixty miles north of Bombay. Maharashtra state police had taken over the construction site of Tarapur Atomic Power Station, then in the throes of a labor strike. The strike had been staged by employees of the American firm Bechtel, the main construction subcontractor on the project. As construction began, Bechtel had recognized a workers' union affiliated with the Indian National Trade Union Congress, a wing of the Congress Party. Dissatisfied with the conservative, nonconfrontational company union, Bechtel employees had organized an alternative union affiliated with the Bombay Labour Union, led by radical labor organizer George Fernandes.[1]

On December 9, 1965, Bechtel employees went on strike to pressure management into recognizing their alternative union. For most of December, the work site was quiet, with Bechtel employees off the job and only the machinery of another subcontractor, Hochtief Modern, chugging away on a cooling water intake canal. But then on Tuesday, December 28, tensions flared as strikers threw stones at other workers who had come to the site to start up some Bechtel machinery so that

[1] "Eight Workers Die in Police Firing," *HT*, December 31, 1965.

W. A. T. Logan, *A Technological History of Cold-War India, 1947–1969*, Palgrave Studies in the History of Science and Technology,
https://doi.org/10.1007/978-3-030-78767-7_9

Fig. 9.1 The main gate of the Tarapur Atomic Power Project site, ca. 1969. The ventilation stack and reactor building are visible in the background (*Source* Photographs of Assistance Programs in Foreign Countries, RG 286-CP, NARA)

it would not seize up out of neglect. Four Americans received minor injuries.[2]

In response to the rising tensions, project management asked the police to come and keep order. Two companies of State Reserve Police deployed to Tarapur to take control of the site. The next day, police and strikers clashed at two locations on the project site, the canteen and the front gate. In both places, police fired tear-gas shells and then live bullets at the strikers. Eight strikers were killed, and dozens injured (Fig. 9.1).[3]

The strike lasted for another month, until at last management agreed to recognize Bombay Labour Union. It was the biggest and most violent confrontation in a project that was fraught with difficulties from the start.

[2] Tarapur Atomic Power Project Quarterly Progress Report No. 5, for October 1, 1965 to December 31, 1965, box 20, USAID Closed Project Loan Files 1960–1971, RG 286, NARA; "Tarapur Strikers Turn Violent," *TOI*, December 29, 1965.

[3] "Eight Die in Police Firing in Tarapur," *TOI*, December 31, 1965.

There were other, lesser labor disputes on the project, as well as a war with Pakistan that disrupted construction. Because of problems unrelated to the labor disputes, commissioning of the station would fall more than a year behind schedule. Then, Tarapur would have teething troubles for years as it started supplying power to the grid on a commercial basis. From the beginning, it would be under a cloud of suspicion for the potential danger it posed to its workers and the environment.[4]

Yet crisis is not the only way to view Tarapur. Tarapur Atomic Power Project has many aspects, and it will not serve to consider only one. Despite the intermittent labor disputes, industrial peace would prevail more often than not. The station did supply power to the grid, and the problems with its commercial commissioning were not always directly related to the station itself. Despite a spate of alarmist articles, Tarapur Atomic Power Station has never had a nuclear accident—which is more than can be said for plants in the United States, Soviet Union, and Japan.

In this chapter, we will take a domestic view of Tarapur plant's construction, commissioning, and early years of operation. Because of the labor disputes in the project, Tarapur offers a chance to study labor in the context of Indian development. Labor is important for development anywhere, especially in large construction projects—and this is nowhere more true than early independence period India, where labor substituted for machinery wherever possible. We will consider the lived experience of workers and management on the project.

This chapter will also consider the technology of the project. Tarapur Atomic Power Station was a foreign technological system adapted—largely by foreign experts—to the Indian environment. Quite unlike the Brahmaputra Bridge (built by two domestic contractors) and Umiam Dam (built departmentally by Assam State Electricity Board), Tarapur was built by General Electric and Bechtel as a turn-key project, which means that the foreign companies were entirely responsible for the project's implementation. Although India played a bigger role than simply receiving the completed station and turning a key to start it running—India's contribution of land, materials, capital, labor, and expertise was in fact considerable—the term "turn-key" vividly expresses the foreign origin of the project's technology.

[4] "Industrial Disputes in the Maharashtra State: Annual Review for the Year 1966," *Labour Gazette*, January 1968, 599–600.

Ultimately, the technology represented by Tarapur Atomic Power Station came to be of only limited use in modern India. In 1973, four years after Tarapur was completed, the first reactor of India's second foreign-designed nuclear powerplant, Rajasthan Atomic Power Station, came online. The reactors of RAPS were Canadian-style CANDU (pressurized heavy-water or PHWR) reactors. CANDU-style reactors rather than American-style light water reactors were used in India's first domestically designed commercial nuclear plant (Madras Atomic Power Station, commissioned 1984), as well as all but one of the Indian nuclear plants built subsequently. In 2005 and 2006, when the third and fourth reactors at Tarapur came online in a new, greatly enlarged plant facility, the new reactors were PHWRs. American-style reactors thus were a technological dead-end in India.[5]

Nuclear power itself has been of only limited use in India. Even in the twenty-first century, the percentage of India's electricity generated by nuclear stations remains in the single-digits—below hydropower and far below fossil fuels. In France, nuclear energy came to supply the majority of the country's electricity and also become an integral part of national identity. In India, by contrast, nuclear power has so far failed to have much influence on either front.[6]

The Makers of Tarapur Atomic Power Station

General Electric won the prime contract for Tarapur Atomic Power Station from the Government of India, but GE did not build the plant alone. The Department of Atomic Energy was there, of course, overseeing every step of the project. There was also a diverse cast of other players—subcontractors, consultants, and auditors, as well as other representatives of the Indian government. The most important subcontractor was Bechtel India Ltd., which GE's Indian branch International General Electric hired in May 1964 to be engineer-constructor of the project.

[5] "Rajasthan Atomic Power Project," *Economic and Political Weekly*, April 13, 1968, 621; "Madras n-Power Unit Goes Critical," *TOI*, July 3, 1983; Power Reactor Information System, "Country Details: India," International Atomic Energy Agency, https://pris.iaea.org/PRIS/CountryStatistics/CountryDetails.aspx?current=IN (accessed February 28, 2020).

[6] Gabrielle Hecht, *The Radiance of France: Nuclear Power and National Identity after World War II* (Cambridge, MA: MIT Press, 1998).

Because of this arrangement, the employees of Bechtel, not GE, did most of the construction work, and Bechtel's own engineers designed the non-nuclear parts of the facility. Bechtel India's parent company was the San Francisco-based engineering and construction firm Bechtel Corporation. By the mid-1960s, Bechtel boasted an impressive résumé of projects that it had built across the United States and around the world, including oil industry infrastructure in the Arabian Peninsula, mines in Venezuela and South Africa, and pipelines in Canada, Austria, France, and Indonesia. Bechtel's experience with nuclear power went back to the world's first functioning power reactor, Experimental Breeder Reactor 1, which first went critical in 1951. Bechtel also served as engineer-constructor for the Dresden Nuclear Power Plant outside Chicago, which featured the first large-scale use of General Electric BWR technology for commercial power production.[7]

An assortment of other subcontractors provided services at the site, including Chicago Bridge & Iron (drywell assembly), Hochtief Modern (construction of the intake and discharge canals), and Blue Star Project (reactor control system). An additional presence at Tarapur was the Kuljian Corporation and its local subsidiary, Kuljian-India. Kuljian had been hired to serve as consultant for the project, as the USAID loan agreement stipulated that the Indian government had to hire its own consultants. This was standard USAID policy for thermal plants. While the Assam State Electricity Board had resisted this requirement for Umiam Dam on the grounds that it had sufficient expertise in hydroelectricity already (see Chapter 7), the Department of Atomic Energy accepted this clause because domestic experience in nuclear energy was limited. Engineers from Kuljian joined the project after receiving an Indian government contract in May 1965.[8]

[7] Tarapur Atomic Power Project Quarterly Progress Report No. 1, for June 19, 1964–December 31, 1964, box 19, USAID Closed Project Loan Files 1960–1971, RG 286, NARA; Richard Finnie, *Bechtel in Arab Lands: A Fifteenth-Year Review of Engineering and Construction Projects* (San Francisco: Bechtel, 1958); Robert L. Ingram, *The Bechtel Story: Seventy Years of Accomplishment in Engineering and Construction* (San Francisco: Bechtel, 1968).

[8] M.N. Chakravarti, "Tarapur Atomic Power Project," *IJPRVD* 18, no. 2 (February 1968): 43; Consulting Engineering Service for Tarapur Atomic Power Project, Progress Report No. 1 for period ending June 30, 1965, box 16, USAID Closed Project Loan Files 1960–1971, RG 286, NARA.

The Kuljian Corporation had been founded in 1930 by Harry Kuljian of Philadelphia. The company worked on engineering projects worldwide, mostly in the electric power sector. Kuljian expanded his company's operations to India shortly after independence, when he met Sadhan Chandra Dutt, an Indian mechanical engineer who had come to the United States to learn about electrical engineering from General Electric. Kuljian trained Dutt for a year in Philadelphia, then made him his permanent representative in India. In August 1961, Kuljian's Indian operation was spun off as its own company, 51 percent Indian-owned, with the remaining 49 percent of shares retained by the parent corporation in Philadelphia. Kuljian's stated intention was to spread the benefits of industrial technology by nativizing engineering expertise.[9]

The project authorities, contractors, and subcontractors on Tarapur Atomic Power Project also hired individual consultants from American universities on an ad hoc basis. Alongside American industry and USAID, American academics were key players in international development during the Cold War. One such academic-consultant was J.W. Johnson, a professor at the University of California-Berkeley. Johnson consulted for engineer-constructors Bechtel, whose main offices were located across the bay in San Francisco. Based on a Bechtel field trip report and model studies performed in Berkeley and Poona, Johnson recommended changes to the design of the cooling water intake system. Johnson and Roy W. Carlson, who installed the instrumentation in Umiam Dam (see Chapter 7), are two examples of American academics who contributed to international development projects. Without the expertise of these academics, large-scale international development projects would not have been possible (Fig. 9.2).[10]

[9] A.K. Ganguly, "U.S. Engineer Helps Develop India's Power Resources," *Indian Railway Gazette*, July 1963, 287–88. In addition to Kuljian Corporation, McLain Associates and Nuclear Utility Services, both of the United States, provided further specialized consulting services.

[10] J.W. Johnson to Burr H. Randolph, December 9, 1963, "Tarapur Nuclear Project, 1963–1964," J.W. Johnson papers, box 1, folder 93, WRCA; R.L. Wiegel et al., "Model Study of Proposed Cooling Water Systems for Nuclear Power Plant to be Located at Tarapur, India" (Berkeley: University of California Hydraulic Engineering Laboratory, December 1964), WRCA; R.L. Wiegel, Yuan Jen, and Ismail Mobarek, "Model Study of Wave Resonance in the Cooling Water Intake Structure, Tarapur Nuclear Power Plant" (Berkeley: University of California Hydraulic Engineering Laboratory, April 1965), WRCA.

Fig. 9.2 One of two huge General Electric transformers installed on either side of the turbine-generator building at Tarapur Atomic Power Station. The shield on the cabinet door is a USAID symbol (*Source* Photographs from the Country Files, RG 286-CF, NARA)

Manufacturers across the contiguous United States contributed parts for Tarapur. Although most of the major suppliers were located in the eastern industrial belt, manufacturers in Tennessee, California, Oregon, and Washington also supplied components. Several different General Electric facilities contributed to the project: the turbine generators came from Schenectady, New York; the reactor control rods were built at the Atomic Power Division in San Jose; the switchgear came from Philadelphia; and the transformers came from Pittsfield, Massachusetts.

As far as possible, project authorities ordered materials from Indian suppliers, in keeping with the government's policy of promoting domestic production. Aggregate was quarried from a riverfront site and screened and washed at the quarry, then transported to the job site and mixed with Indian cement to form the concrete that made the majority of the structure. Reinforcing steel came from India's growing steel industry. In general, Indian manufacturers supplied less capital-intensive materials, in keeping with what Indian industry was able to produce at that time. The most sophisticated components supplied by Indian industry were the control and instrument panels of the main control room, built by Atomic Energy Establishment, Trombay.[11]

Work at Tarapur

The construction of Tarapur Atomic Power station began in September 1964 after the end of the monsoon. Before construction proper could begin, though, the project site and surrounding area had to be cleared and made accessible. At the Boisar railhead, two warehouses were built to store materials until they could be transported to the site. To access the site, workers built a 2,000-foot approach road, branching off from the Boisar-Tarapur Road. They also dug tube-wells to provide fresh water. The Maharashtra State Electricity Board strung a branch line to provide power from its grid. The entire site was leveled to stand at a regular elevation of 103 feet, relative to the project's datum. The datum did not represent height above mean sea level; instead, it was based on a preexisting structure, the plinth of Tarapur lighthouse, which was considered to stand at 100 feet. Low tide could be as low as 74.5 feet on the project datum, or 25.5 feet below the lighthouse plinth. Zero on the datum was

[11] TAPP quarterly report 4, July–September 1965, box 20, USAID Closed Project Loan Files 1960–1971, RG 286, NARA.

always deep underwater. For the duration of the construction, progress reports gave building heights based on the datum, where grade stood at 103 feet. Thus the roof of the reactor building, given in the reports as being at elevation 248′ 6″, actually stood at 145′ 6″ above grade, or about 165 feet above mean sea level.[12]

Excavation at the project site began on September 13, 1964, with blasting beginning three months later. In January 1965, the first concrete for the permanent plant buildings was poured. By the end of April, the plant buildings were far enough along that Chicago Bridge & Iron could start erecting the drywells—two huge steel spheres, each 65 feet in diameter, that would house the station's twin reactors. The drywells had been fabricated in pieces in Greenville, Pennsylvania and were assembled on-site with the help of a towering derrick crane standing on the west wall of the reactor building (Fig. 9.3).[13]

In April 1966, the first reactor core arrived at the project site on board a specialized landing craft transport, *Inagua Gull*. The reactor vessel, which was over fifty feet long and weighed 268 tons, was too bulky to transport by rail from Bombay. Instead, the *Inagua Gull* beached itself on a purpose-built landing spit at the project site. The bow of the ship split and opened like a clamshell, and a ramp lowered onto the beach. Work crews used machinery and a fair bit of ingenuity and elbow grease to transfer the reactor to a wooden sled and haul it across the site to the reactor building. On May 29, the crane that had been used to assemble the drywells lifted the reactor vessel off the ground and lowered it into its home inside the drywell. Later that year, the second reactor arrived from the United States and was installed in a similar manner.[14]

[12] TAPP quarterly report 1; "Tarapur Reporter," August 5, 1966, box 24, USAID Closed Project Loan Files 1960–1971, RG 286, NARA (hereafter cited "TR" and the date); "Meeting with consultant to discuss status of circulating water system design, San Francisco, February 17, 1964," "Tarapur Nuclear Project, 1963–1964," J.W. Johnson papers, box 1, folder 93, WRCA.

[13] TAPP quarterly report 1; Tarapur Atomic Power Project Quarterly Progress Report No. 2, for January 1, 1965–March 31, 1965, box 20, USAID Closed Project Loan Files 1960–1971, RG 286, NARA; Tarapur Atomic Power Project Quarterly Progress Report No. 5, for October 1, 1965 to December 31, 1965, ibid.; "Tarapur Reporter," June 24, 1966, box 24, USAID Closed Project Loan Files 1960–1971, RG 286, NARA.

[14] GE, TAPS monthly report 23, April 1966, box 19, ibid.; GE, TAPS monthly report 24, May 1966, ibid.; USIS, "Nuclear Power from Tarapur," 1969, US Moving Images compiled 1982–1999, RG 306, NARA.

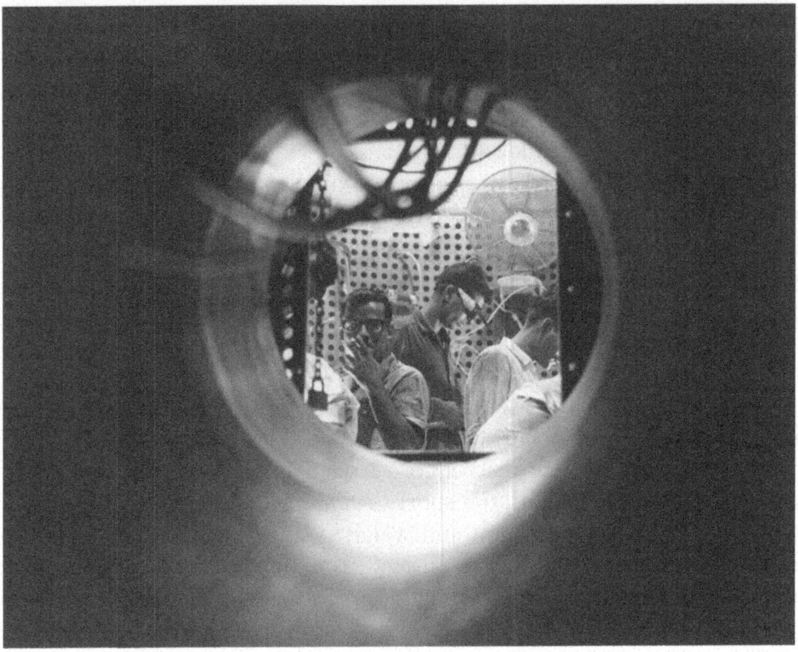

Fig. 9.3 Indian technicians working inside one of the TAPS reactors (*Source* Photographs from the Country Files, RG 286-CF, NARA)

The Tarapur project included more than just the main plant building with the reactors, turbine generators, and control room. Dominating the site, a 366-foot ventilation stack stood adjacent to the southeast corner of the reactor building. The stack vented air out of the reactor building. Despite multiple layers of shielding, it was possible that the vented air could contain small levels of radiation. The height of the stack was intended to give the radioactive material a chance to diffuse to harmless levels before it reached the ground. Venting only took place when a weather station mounted on a nearby 650-foot tower indicated that atmospheric conditions were favorable for dispersal.[15]

[15] P. Abraham, "Safety Features at Tarapur," *The Indian and Eastern Engineer*, February 1970, 117–20; V. Ramachandran, et al., "Design and Construction of the 120

Next to the stack stood a 600,000-gallon tank for storing the freshwater used by the reactors in their boiling water cycle. The water came from Sakhare Reservoir, impounded by a purpose-built dam fifteen miles northeast of the plant. Buried pipes carried the reservoir water to the plant site. (Other users of the reservoir water included Western Railway steam locomotives and 29 nearby villages.) The Indian project officials recommended adding the storage tank to the plant in the event that the water mains were disrupted.[16] Also on the site were a radwaste building for storage of spent nuclear fuel; an administration building with a conference room and a cafeteria; a workshop and warehouse building; a switching yard that linked the Gujarat and Maharashtra grids (built by the Maharashtra State Electricity Board); and a system of intake and discharge canals for cooling water, which included two 3,000-foot jetties stretching out into the Arabian Sea (Fig. 9.4).

The construction relied heavily on imported machinery. The largest of these was the 300-ton stiff-leg derrick crane used for handling the reactor cores and other particularly cumbersome pieces of equipment. On a similarly large scale, the intake canal, built by subcontractor Hochtief Modern, was excavated with the help of two gantries that moved along rails laid on the seafloor. Riding on the back of the gantry, a backhoe scooped up the seafloor between the rails. On a smaller scale, but no less important for the project, were American trucks, jeeps, and other smaller construction machines for digging, lifting, and moving. According to the first TAPP quarterly progress report, thirty-seven shipments left the United States bound for the project in the latter half of 1964, consisting mostly of construction equipment "including dump trucks, rear dumpers, traxcavators, tractor dozers, mobile cranes, crawler cranes, dragline buckets, fork lifts, [a] sand washing plant, [a] concrete batching plant, stone crushers, trucks, and welding sets."[17]

Metre High Meteorological Tower Facility at the Tarapur Atomic Power Station," *The Journal of the Institution of Engineers (India)* 48 (September 1968), 268–69.

[16] M.N. Chakravarti, "Tarapur Atomic Power Project," *IJPRVD* 18, no. 2 (February 1968): 46–47; TAPP quarterly report 1; "Tarapur Water Supply Scheme," *TOI*, April 9, 1969.

[17] TAPP quarterly report 1, June–December 1964, box 19, USAID Closed Project Loan Files 1960–1971, RG 286, NARA; TAPP quarterly report 12, July–September 1967, box 20, ibid.; TR, July 1, 1966. "Tarapur Reporter" excitedly announced the arrival of "BACKHOE" by writing the name in all-caps.

Fig. 9.4 View of the cooling water intake structure on the west side of the plant (*Source* Photographs of Assistance Programs in Foreign Countries, RG 286-CP, NARA)

Just as essential to the project as the mechanized equipment was human labor. A portion of the channel of the intake canal was excavated by hand behind a cofferdam. Although the project reports do not spend much time discussing the non-mechanized aspects of the construction, project photographs and films prominently feature individual men and women trudging across the work site while balancing trays full of cement or aggregate on their heads.[18]

The Tarapur workforce expanded and contracted based on project needs and the status of labor disputes that simmered through 1965 and 1966. At the beginning of construction in the latter half of 1964, the workforce numbered just over 2,000. As construction progressed from sinking foundations to erecting the buildings, the workforce steadily expanded, reaching its peak in first quarter 1966 at just over 6,800. From there, the workforce shrank, although in fourth quarter 1966 it jumped again because of increased demand for welders. By the end of construction in second quarter 1969, the workforce had shrunk to just under 1,000.[19]

Workers on the project came from all over India. According to figures printed in the project newsletter, 3,364 Indian nationals were employed on the project on the last day of June 1965. These workers came from thirteen different states. Maharashtra was dominant, with 1,526 workers. Punjab came in a distant second, with 504 workers. Others came from as far away as Kerala, Andhra Pradesh, and West Bengal. Sixty-three Nepalese were also employed on the project.[20]

Throughout the project, Indians vastly outnumbered expatriates from the United States or other countries. The highest number of American workers on the project, reported in first and second quarters 1967, was seventy-three. During the early phases of construction, the American personnel were predominantly Bechtel employees, with the ranks of GE supervisors remaining in single-digits or the low teens. Later, when concrete construction was mostly complete, the Bechtel force shrank and GE employees increased in number to oversee the installation and testing of the plant's electrical equipment.

The plant had to be built with American expertise, but upon commissioning it would be turned over to complete Indian control. Before this

[18] General Electric, Tarapur Atomic Power Station Progress Report no. 20, January 1966, box 19, USAID Closed Project Loan Files 1960–1971, RG 286, NARA.

[19] TAPP quarterly report 1, for June–December 1964, box 19, USAID Closed Project Loan Files 1960–1971, RG 286, NARA; TAPP quarterly report 6, January–March 1966, box 20, ibid.; TAPP quarterly report 9, October–December 1966, ibid.; TAPP quarterly report 19, April–June 1969, ibid.

[20] TR, July 16, 1965. The next issue of the newsletter reported that the newsletter staff had received complaints for not including the territory of Goa in the list of workers' home states. TR, July 23, 1965.

could happen, an Indian workforce had to be recruited and trained to operate the plant and fix the little or big problems that would crop up in its day-to-day operation. These people would belong to the Operation and Maintenance Group, and project authorities recruited them from places in India where they would have already gained some expertise that would be useful, such as the Atomic Energy Establishment Trombay, electric utilities such as the Damodar Valley Corporation, and Tarapur Atomic Power Project itself.[21]

Training for the recruits consisted of both intensive classroom instruction and on-site experience at an operating power plant. On the theoretical side, trainees took a three-week BWR Indoctrination Course and a five-week Tarapur Technology Training Course in San Jose. They also wrote operating manuals for the station.[22]

On the practical side, trainees got experience at conventional power stations in India as well as on the CIRUS reactor at Trombay. A group of six trainees spent five months at Garigliano Nuclear Power Plant, a 150-MW BWR plant in Italy. Other groups trained on the Dresden Nuclear Power Plant outside Chicago or the small 5-MW Experimental Superheat Reactor at Vallecitos near San Francisco. In addition, individuals or groups of two were seconded for longer periods to research labs or industrial facilities to get more in-depth experience. Two TAPP trainees spent a year at Oak Ridge National Laboratory, learning about reactor operations and potential hazards. Others spent time at Honeywell Brown, the GE manufacturing shops, or the National Reactor Testing Station in Idaho Falls.[23]

Upon completing the theoretical and practical training program, recruits took a test similar to the US Atomic Energy Commission licensing test for reactor operators. Project management reported that the results of these tests were entirely satisfactory. Tarapur Atomic Power Station now had a workforce ready to operate it once it was complete (Fig. 9.5).[24]

[21] M.N. Chakravarti, "Training Programme for Operation and Maintenance Personnel of Tarapur Atomic Power Project," June 1966, box 24, USAID Closed Project Loan Files 1960–1971, RG 286, NARA; M. Dayal, "Training of Operation and Maintenance Staff for the Tarapur Atomic Power Station," July 1967, ibid.

[22] Ibid.

[23] Ibid.

[24] Ibid.

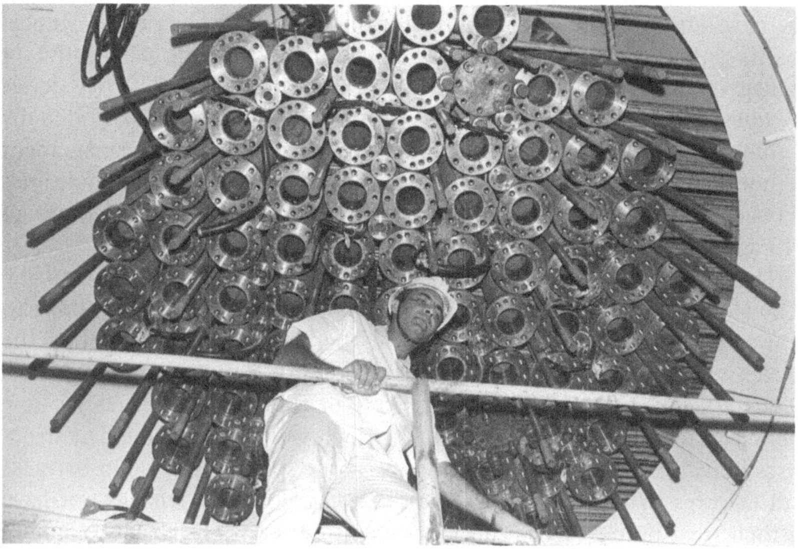

Fig. 9.5 Maheshwar Dayal, project engineer, standing below the control rods of one of the reactors at Tarapur Atomic Power Station (*Source* Photographs from the Country Files, RG 286-CF, NARA)

Life at Tarapur

Tarapur was not just a site of labor, it was also a community, which at any given time consisted of thousands of Indians and upwards of a hundred Americans employed on the project, as well as the families of both. More precisely, the Tarapur workforce consisted of several different communities, divided by class, ethnicity, race, language, and occupation. Sometimes these communities interacted peacefully, but occasionally, in the case of the 1965–66 strike, they clashed violently.

The most complete source in the official records about community life at Tarapur Atomic Power Project is the official weekly newsletter, "Tarapur Reporter." Content of the newsletter included project progress reports, profiles of the different divisions on the project, world news digests, community news and classifieds, riddles and puzzles, and safety cartoons. The newsletter was initially issued exclusively in English from the project's Bombay engineering office, with Hindi and Marathi editions appearing later. As it provides details not found in the official reports, the newsletter is an essential source for understanding the project. It also has its pitfalls. "Tarapur Reporter" was written from the perspective of management; the workers' perspective is not represented. The newsletter portrays the project workforce as a harmonious community united by a common objective, but the conflicts between labor and management demonstrate that the workforce was far from harmonious. The fault lines between workforce communities are evident from a close reading of "Tarapur Reporter."[25]

The most distinct dividing line separated the National Camp and the IGE Colony. The National Camp, adjacent to the jobsite, was built to house the Indian laborers working on the project. The camp had the capacity to house about 2,150 men, or roughly half of the labor force. (The remainder found accommodations in the nearby villages.) These men lived in sixty temporary barracks, which were some of the first structures built for the project. Although certainly spartan, the camp barracks had features that many of these working-class Indian men did not have in their own homes. Each barrack was outfitted with electricity, running water, and kerosene. In addition, the common areas of the camp had such recreational facilities as a football field, a volleyball court, and a recreation hall with a library and a reading room. A permanent market was located just outside the grounds of the camp. National Camp itself had a staff of 86 people. The diversity of the inhabitants of the National Camp led to its receiving the nickname of "Chota India" or "Little India" (Fig. 9.6).[26]

Seven miles away, on the Boisar-Tarapur Road, was the IGE Colony, a permanent housing development. The Indian and American administrators and engineers overseeing the project lived in the colony with their families. The colony had five grades of bungalows, with status on the project reflected by the size of one's bungalow. Expatriates working on the project only for a short term stayed in single-occupancy accommodations and ate at the colony's mess hall. In January 1965, when the colony's buildings were three-quarters complete, eight expatriate and seventeen Indian families had already taken up residence. By August

[25] Partial runs of the English edition of "Tarapur Reporter" from 1965 and 1966 are found in box 24, USAID Closed Project Loan Files 1960–1971, RG 286, NARA. The author is grateful to Arne Kaijser for recommending a critical interpretation of "Tarapur Reporter."

[26] "The Men Behind the Job: National Camp and IGE Colony," TR, August 5, 1966; TR, August 12, 1966.

Fig. 9.6 Don T. Spangenberg and G.J. Larsen of General Electric in the Tarapur control room (*Source* Photographs of Assistance Programs in Foreign Countries, RG 286-CP, NARA)

1966, the total number of families had increased to 104, and they had been joined by 40 expats living without their families.[27]

The IGE Colony had many modern comforts that were inaccessible to the majority of the Indian population—including American appliances and consumables imported under an omnibus license that also covered materials for the construction project. Despite these comforts, living in the colony was not the same as living in an American suburb. Significantly, the colony was far more integrated than the average American neighborhood. Even though they did not fit neatly into America's black-and-white racial system of the 1960s, Indians in the United States during this time suffered discrimination and marginalization on account of their dark skin color. Some had trouble renting or buying property.

[27] GE, TAPS monthly report 8, January 1965, box 21, USAID Closed Project Loan Files 1960–1971, RG 286, NARA; TR, August 5, 1966; Arun Joshi, interview by author, Kennewick, WA, October 5, 2017.

De-facto white suburbs often excluded Indian immigrants as well as African-Americans. The IGE Colony, by contrast, was not segregated. Both Indian and American families lived in the colony, with Indians constituting the majority.[28]

Both the National Camp and the IGE Colony hosted special events on a regular basis. In both places, films in Hindi, Marathi, and English were screened once or twice a week. Titles screened included *Abbott and Costello Lost in Alaska* (1952), *The Millionairess* (1960), and the Raj Kapoor romantic comedy *Dulha Dulhan* (1964). At the National Camp, football matches and table-tennis tournaments attracted participants from both residential areas. These events, along with the amenities provided at the residential areas, are expressions of industrial paternalism—a management style that entailed giving workers certain benefits in order to keep them happy or in any event disinterested in unionization. Industrial paternalism was an American progressive-era response to the exploitation and labor strife of the Gilded Age. It also caught on early in some industries in India, notably Tata Iron and Steel Company. At Tarapur Atomic Power Project, with its film screenings, sporting events, recreational facilities, and even a company union (Tarapur Anushakti Kamgar Sangh), industrial paternalism was in full flower.[29]

As another expression of industrial paternalism, management hosted events to mark important days in both the Indian and American calendars. During the monsoon season, Indians and Americans celebrated their respective national holidays. According to "Tarapur Reporter," the residents of the IGE Colony celebrated American Independence Day with a potluck dinner at the colony clubhouse on July 3, 1966, and a fireworks display launched from the baseball field the following evening. The next month, Tarapur Atomic Power Project marked Indian Independence Day, August 15, with a major celebration. On the evening of August 14, a variety show featuring entertainers brought in from Bombay played to a packed auditorium in the National Camp. On the morning of August 15, Burns Woodward, the acting GE manager of the project, performed the

[28] TAPP quarterly report 1; Andrew J. Rotter, *Comrades at Odds: The United States and India, 1947–1964* (Ithaca, NY: Cornell University Press, 2000), 166–67.

[29] TR, June 24, 1966; TR, July 8, 1966. For American progressive-era industrial paternalism, see Andrea Tone, *The Business of Benevolence: Industrial Paternalism in Progressive America* (Ithaca, NY: Cornell University Press, 1997). For industrial paternalism's manifestation in India, see Andrew Sanchez, "Deadwood and Paternalism: Rationalizing Casual Labour in a Company Town," *Journal of the Royal Anthropological Institute* 18, no. 4 (December 2012): 808–27.

traditional Indian flag-raising on the football ground, which was followed by a march-past of the project's security force. Then the project hosted an enormous open house, reportedly visited by 40,000 people who had traveled there from the local area as well as from as far away as Bombay.[30]

The Tarapur communities also participated in religious festivals and services. In the IGE Colony, the Tarapur Union Church met every other Sunday. A dedicated church building had not been included in the colony plan, so the congregation met in the clubhouse or the schoolroom. Christian members of the Tarapur workforce took communion at the church and even christened their children there.[31]

Far outnumbering the Christians, the Hindus in the workforce also performed their own religious rituals at the site. Arun Joshi, a Poona-born telecommunication and instrumentation engineer on the project, organized the celebration of Ganapati (Ganesh Chaturthi), a festival that falls in August or September and is especially popular in Maharashtra. The Tarapur workers observed Ganapati by setting up a four-foot clay idol of the god Ganesh on a platform in the National Camp recreation hall. For ten days, the idol stood on the platform, and thousands of workers and their families, as well as residents of nearby villages, came to perform *puja* (worship), sing *bhajans* (devotional songs), and receive *prasad* (sweets). At the end of the ten days, the worshipers carried the idol down to the beach and immersed it in the ocean waves adjacent to the project site.[32]

At about the same time as Ganapati, the workers also celebrated Vishwakarma Puja. This was a significant festival for the workers on Tarapur Atomic Power Project because it is a time when craftsmen revere the tools that they use for their work. In preparation for the festival, the workers decorated all the project offices with banana plants, palm leaves, and flowers. The workers, along with their families, sang *bhajans* and performed traditional dances.[33]

The residential segregation of National Camp and IGE Colony reflected a hierarchical division of the workforce based on class. Another demonstration of class-based hierarchy was the Tarapur Swim and

[30] "Fourth of July Celebrated," TR, July 8, 1966; "Thousands Visit Tarapur Atomic Power Project," TR, August 19, 1966.

[31] TR, June 17, 1966; TR, July 22, 1966; TR, August 12, 1966.

[32] TR, September 9, 1966; TR, September 23, 1966; Arun Joshi, interview by author, Kennewick, WA, December 8, 2017.

[33] TR, September 23, 1966.

Racquet Club. Established in September 1965 in IGE Colony, the club was meant to replicate the functions of the elite social clubs of Bombay and other Indian cities by providing its members with recreational facilities and regular social activities. The club's facilities included a swimming pool, two tennis courts, an air-conditioned badminton court, a library, and a lounge with a piano and a radiogram (a combined radio and record player). The club hosted monthly parties and dances for its members, usually corresponding to western holidays such as Valentine's Day and St. Patrick's Day. By the first anniversary of its founding, the club had become such an integral part of IGE Colony life that "Tarapur Reporter" declared it to be the "centre of the social activities in the colony."[34]

Social clubs in India, like so many other institutions, originated in the colonial period to serve the needs and wants of British colonialists. Some British clubs had racially exclusive membership and admission policies, but complete racial exclusion was the exception rather than the rule, with only four percent of clubs holding such policies. Most clubs allowed some degree of participation by Indians, and over time the clubs were Indianized by admitting certain Indians as patrons (usually members of the princely class), honorary members, and finally full voting members. After independence, the remaining exclusive clubs either closed their doors or adopted more inclusive membership policies—sometimes under external pressure such as the threat of government expropriation of clubland or the denial of liquor licenses for clubs with no Indian members. The last all-British club in Bombay, the Bombay Club, closed in 1959. By the end of the 1960s, membership of the formerly exclusive clubs was mostly Indian. Upper-class Indians had thus replaced Europeans in the social clubs, just as they slotted into the hierarchy of modern India as a whole after the departure of the British. As a racially inclusive but class-selective organization, the Tarapur Swim and Racquet Club resembled other social clubs of 1960s India.[35]

The Tarapur workforce and communities were divided on lines of class if not race. But what about gender? Typically for a large construction project, the management was all male, as well as most of the workforce. National Camp provided accommodations for male laborers only. Sexism

[34] TR, December 10, 1965; TR, September 9, 1966, TR, September 23, 1966.

[35] Benjamin B. Cohen, *In the Club: Associational Life in Colonial South Asia* (Manchester: Manchester University Press, 2015), 122–23, 169–78.

appears to have been normalized on the project to some extent, as pin-up girl calendars were distributed on the site with official management blessing. "Tarapur Reporter" announced the calendars in a self-censored blurb in December 1965: "A single sheet calendar illustrating a pretty, vivacious .... (you know what) [*sic*] is now under print, and shall be ready for distribution by the end of this month." The newsletter went on to say that (presumably American) GE project officials had been involved in the selection of the pin-up, and one of them called her "An Atomic Beauty."[36]

But Tarapur Atomic Power project was not an exclusively male undertaking either. Women served alongside men as manual laborers on TAPP, which is not uncommon on construction projects in India. (Although almost never mentioned explicitly in project documentation, these women laborers do appear in photographs and films taken at the site.) On the administrative side, the publisher of the English-language edition of "Tarapur Reporter," Leela Sajnani, was a woman, along with some of the contributors and assistants in the copying and stapling of the newsletter. Some of the wives of project engineers had jobs peripheral to the project, such as communication engineer Arun Joshi's wife, who taught in the school for project children. Other wives, with no formalized role in the project, performed domestic duties of cooking, cleaning, and child-rearing in the IGE Colony. They also did their share of socializing. When Mrs. Root, the wife of project manager M.D. Root, returned to Tarapur after three and a half months in the United States, the other project wives welcomed her back with a ladies' coffee at the club.[37]

In our survey of the people groups of Tarapur Atomic Power Project, let us not overlook the youngest expatriates: the American children whose families had temporarily relocated to India for the project. For many, this was not their first time living outside the United States, as their fathers were self-described "industrial gypsies" who moved from one international megaproject to the next.[38] But whether this was their first time abroad or they had never really lived in the United States, in India these

[36] TR, December 10, 1965.

[37] TR, July 16, 1965; TR, July 23, 1965; TR, September 23, 1966; Arun Joshi, interview by author, Kennewick, WA, October 5, 2017. One of the few textual sources that mentions women's contribution of labor to the project is "Tarapur: Gateway to the Nuclear Age," *Participant Journal*, January 1970, 12–14.

[38] The term "industrial gypsy" comes from a conversation my father had with a Bechtel employee while working in San Francisco in the 1980s.

children were strangers in a strange land, and they had their own distinctive experience there—different from their Indian counterparts but also different from their predecessors, the British children who had lived in India during the Raj.

British colonialists had structured their lives based on the duties of their class and race. Children lived in India from birth to about the age of five, when their parents sent them to England for their schooling, where they stayed there until they were grown and ready to return and take up the work of empire.[39] For American children at Tarapur, there was no set structure for their lives. Unlike both British and Indian cultures, American culture does not have a strong boarding school tradition. The American children stayed with their families and attended ad hoc schools set up in the IGE Colony to educate the children of project employees.

The teachers at the project school were Indian-born women who had experience studying and teaching in India, the United States, and several countries in Europe. One of them, Mrs. Banudwala, had prior teaching experience at another industrial site, a chemical plant in Uttar Pradesh. For older children, a corner of the Bombay engineering office was set aside for high school classes. An Indian teacher, M.K. Chitanvis, taught a handful of expat teenagers using correspondence course material produced by the University of California.[40]

Much as the adults had their sporting events, the project children had their own opportunities for structured fun, provided by authority figures. Toward the end of the 1966 monsoon, children from first through third grades in the colony school went on a picnic and field trip to nearby attractions. With Labh Singh acting as their driver and guide, these six children explored the ruins of Tarapur Fort, visited a Shiva temple, and toured a fishing village, before eating their lunch on Chinchani Beach and heading back to the IGE Colony.[41]

Especially for those expat children who had never experienced the tropics before, living in India offered a spate of novel experiences. Some could be frightening or confusing, such as the variety of tropical insects and snakes they would encounter while playing outside or even in their

[39] Charles Allen and Michael Mason, *Plain Tales from the Raj: Images of British India in the Twentieth Century* (New York: St. Martin's Press, 1976), 31–33.

[40] "Tarapur Teachers," TR, July 23, 1965; "High School," TR, December 10, 1965.

[41] "School Picnic," TR, August 5, 1966.

own houses. The monsoon was also a singular experience. It struck Tarapur with force, sometimes raining almost continuously for a week at a time. "Although construction work was delayed, the kids had fun with puddles of rainwater and overflowing storm drains," the project newsletter reported in July 1965. "Some were luckier than others. They could sail paper boats in their bedrooms."[42]

As for children of the laborers on the project, it is difficult to say much about them. Children of men living in National Camp stayed in their home villages, while some of the men living in the neighboring villages were able to bring their families with them. If these children went on any organized excursions, this was not reported in the project newsletter. Indeed, "Tarapur Reporter" had much less to say about the laborers than about the residents of IGE Colony.

Although the Tarapur workforce was divided on lines of class or status, with the managers and engineers enjoying better living conditions and more amenities than the laborers, this division did not lead inevitably to unrest by the workers and a violent reaction by management and police. The strike and its reaction have their own historical context in which they must be understood, and it is to this that we turn next.

Labor and the Indian State

Working-class organization and dissent have a long history in western India, almost as long as mechanized industry itself. India's first modern factory, a textile mill, opened in Bombay in 1854. Although textile workers did not formally organize until 1926, they started staging strikes as early as the 1880s. The first Indian workers' union to be organized was the Amalgamated Society of Railway Servants, founded in 1887. In 1920, the All-India Trade Union Congress (AITUC) became India's first national labor organization. From this point until 1947, trade unions were closely associated with the Congress-led nationalist movement. Mohandas Gandhi helped organize Ahmedabad textile workers, and Jawaharlal Nehru served as president of AITUC in 1929.[43]

[42] TR, July 23, 1965.

[43] Morris David Morris, "Labor Discipline, Trade-Unions, and the State in India," *Journal of Political Economy* 63, no. 4 (August 1955): 294–96; Mrinal Datta Chaudhuri, "Labor Markets as Social Institutions in India," *Iris-India Working Paper* 10 (March 1996), 6–7, http://pdf.usaid.gov/pdf_docs/Pnaby353.pdf.

AITUC remained the only national trade union organization for twenty-seven years, until the national political field got more complicated at independence in 1947. When socialists and communists left Congress to start their own political parties, organized labor also fragmented on ideological lines. AITUC had been controlled by communists since 1927, so non-communist labor leaders left the organization in the year of independence and established the Indian National Trade Union Congress (INTUC). Allied with the Congress Party, INTUC would be the most conservative of the national labor organizations. The socialists also established their own national labor organization, the Hind Mazdoor Sabha (HMS). These three organizations—AITUC, INTUC, HMS—would be political instruments of the parties that controlled them. Labor struggles in the factories and on the streets were directly linked with the party politics played out in parliament and the state legislative assemblies. Mrinal Datta Chaudhuri estimates that 20–30 percent of labor disputes in the period 1950–1975 were conflicts for union recognition or rivalries between unions.[44]

In the late fifties, a younger, more militant generation of activist leaders appeared on the Bombay labor scene. These leaders, mostly radical socialists, established the Bombay Labour Union in July 1959. As its name suggested, the Bombay Labour Union was meant to be comprehensive, providing support and a voice for workers in industries that were too small or fragmented to form viable unions of their own. Union organizers began recruiting hotel workers, barbers, laborers at small engineering works, and workers at private hospitals. The union also took over some preexisting communist or socialist unions. By April 1960, less than a year after its inception, the Bombay Labour Union boasted more than 25,000 members.[45]

The leader of the new guard of labor was George Fernandes. Only twenty-nine when he became president of the newly founded Bombay Labour Union, Fernandes already had nearly a decade of experience in the labor movement, having moved to Bombay in his early twenties after quitting studies to become a Catholic priest. Throughout the 1960s, Fernandes made himself the bane of the municipal authorities, staging

[44] Ralph James, "Politics and Trade Unions in India," *Far Eastern Survey* 27, no. 3 (March 1958): 41; Datta Chaudhuri, "Labor Markets as Social Institutions in India," 7.

[45] "The Week at Home: New Trade Union Leadership," *TOI*, April 28, 1960.

demonstrations and mass strikes. He was arrested and kept in preventative detention on several occasions. In July 1960, police arrested him for bodily blocking a train at Dadar station as part of a city-wide *bandh*. In April 1963, he was arrested again and detained this time for eight months for having allegedly taken part in "anti-national activities." In response to his detention, tens of thousands of city employees, public transit workers, taxi drivers, and hotel workers went on strike.[46]

The Indian government had the power to arrest dissidents like George Fernandes and suppress dissent at Tarapur Atomic Power Project because the leaders of independent India had laid claim to and expanded upon the coercive powers of the colonial Government of India. As early as 1860, Section 141 of the Indian Penal Code had defined unlawful assembly as five or more people who had the intention of overpowering the government, resisting the law, committing criminal acts, illegally seizing property, or forcing someone to do something illegal. This law was applied against the nationalist movements of the twentieth century, and it remained in force after independence.[47]

During the world wars, the government granted itself extraordinary powers, ostensibly for maintaining internal security during a period of emergency. After World War I, the government extended the extraordinary powers with the Anarchical and Revolutionary Crimes Act. Although the act soon lapsed, it set a precedent of enacting draconian measures to suppress domestic dissent in peacetime. When civil disobedience expanded in the early thirties, the government passed the Emergency Powers Ordinance of 1932, targeted at nationalist agitation. Congress was banned and thousands of nationalists held without trial in preventative detention. In 1937, Congress was legal again and Congress-led governments were elected in some provinces of British India. In short order, these governments were employing the very same draconian measures that had been used to jail Congress leaders just five years earlier. Not only did repressive measures begin in the colonial era; the use of these

[46] "The Week at Home: New Trade Union Leadership," *TOI*, April 28, 1960; "Sabotage Cases in Bombay Reported," *TOI*, July 14, 1960; "Fernandes Is Arrested," *TOI*, April 6, 1963; "Life in Bombay City Disrupted by Strike," *TOI*, August 20, 1963.

[47] W. Morgan and A.G. MacPherson, *The Indian Penal Code (Act XLV of 1860) with Notes* (Calcutta: G.C. Hay, 1863).

measures by Indians against other Indians began before independence as well.[48]

With the transfer of sovereignty in 1947, India's new rulers wasted no time in adopting and elaborating on the powers of the colonial government. The Indian Constitution included provisions for emergency powers that resembled the emergency provisions of the Government of India Act (1935). Just weeks after the Constitution came into effect in January 1950, Parliament passed the Preventative Detention Act, which allowed the government to hold dissidents without charge for up to twelve months. The act was proposed as a temporary measure, expiring after a year, but Parliament renewed it again and again until finally allowing it to lapse in 1970. Only a year later, preventative detention returned with the Maintenance of Internal Security Act. In a similar fashion, wartime emergency acts that gave broad powers to the executive lasted far longer than any of India's wars. The Defence of India Act 1962, passed in response to the month-long Sino-Indian War, stayed in effect for six years.[49]

If the government of independent India had taken over the role of the coercive state from the British, it had also inherited a tumultuous domestic situation. The mass demonstrations, marches, strikes, hunger strikes, and *bandhs* that had taken place before independence continued after the nationalist cause had been won. Even though the British were gone and India was for Indians, inequality persisted, as did questions of language, religion, caste, and regional or tribal identity.[50]

Just as dissent carried over into independence, albeit with somewhat changed motivations, so too did post-1947 police practices echo colonial responses to nationalist agitations. Police continued to disperse crowds with *lathi* (billy club) charges and sometimes resorted to formation firings into the crowd. The police also adopted newer crowd-control innovations like tear gas or sickening gas (adamsite). Police activities were reported, and often sensationalized, in the press, leading to counter-demonstrations

[48] Anil Kalhan, et al., "Colonial Continuities: Human Rights, Terrorism, and Security Laws in India," *Columbia Journal of Asian Law* 20 (2006): 128–31.

[49] Kalhan, et al., "Colonial Continuities," 132–35; Kuldeep Mathur, "The State and the Use of Coercive Power in India," *Asian Survey* 32, no. 4 (April 1992): 342.

[50] David H. Bayley, "The Pedagogy of Democracy: Coercive Public Protest in India," *The American Political Science Review* 56, no. 3 (September 1962): 663.

and calls by members of opposition parties for retribution against the police.[51]

The Tarapur Strike

The Tarapur strike of December 1965 and January 1966 was a part of the struggle between conservative Congress-supported labor organizations and the militant new guard of labor, epitomized by the Bombay Labour Union. Early in the project, Bechtel management had granted recognition to a company union, Tarapur Anushakti Kamgar Sangh (Tarapur Atomic Power Workers' Union), affiliated with INTUC and led by Jagdish Narain Choubey. (This was another aspect of project management's policy of industrial paternalism, discussed earlier.) When the fight between conservative and radical labor organizations reached the Tarapur project, Bechtel employees started to agitate for recognition of an alternative union affiliated with the Bombay Labour Union. Labor leaders notified Bechtel management that they would strike if their union was not recognized. Since management already recognized the INTUC union, they considered the demand illegitimate and did not respond. As threatened, 4,493 Bechtel employees walked off the job on December 9.[52]

The project site was quiet for several weeks as Bechtel workers stayed off the job, but then the tension between labor and management erupted into violence in the last days of 1965. On Tuesday, December 28, striking workers threw rocks at other Bechtel employees who had come onto the job site to perform preventative maintenance on heavy machinery. Four Americans and a dozen Indians received injuries from the flying stones. (Plenty of windows were shattered as well.) At 3:00 the following afternoon, project management turned control of the site over to the state police. In a series of violent confrontations on Wednesday evening between police and strikers, eight people were killed and more than fifty wounded, one of whom would die a week later in a hospital.[53]

[51] David H. Bayley, "The Police and Political Order in India," *Asian Survey* 23, no. 4 (April 1983): 493.

[52] "Industrial Disputes in the Maharashtra State: Annual Review for the Year 1966," *Labour Gazette*, January 1968, 599–600; "Strike by Tarapur Project Workers," *TOI*, December 24, 1965.

[53] TAPP quarterly report 5, October–December 1965, box 20, USAID Closed Project Loan Files 1960–1971, RG 286, NARA; GE, TAPS Progress Report 19, December 1965,

In a statement released the day after the December 29 shooting, police inspector general S. Majeedulla claimed that his officers had fired in self-defense. In the confrontation with strikers, 52 police were injured, and three remained in critical condition. Majeedulla explained that an angry mob of workers had descended on the project canteen and eaten up all the food, then proceeded to wreck the place in lieu of paying. The police attempted to intervene, but the workers started stoning them, just as they had done to the Bechtel employees trying to perform preventative maintenance on machinery the day before. In response, the police staged a lathi-charge, released tear gas, and fired 52 rounds with their rifles. A second confrontation took place near the project gate. By the inspector general's account, workers set fire to a police jeep and tried to blow up a magazine of blasting explosives, but the police intervened with tear gas and stopped them.[54]

Representatives of the Bombay Labour Union had their own version of events, which contradicted the police account on almost every point. A union representative claimed that workers could not possibly have eaten food at the canteen and then refused to pay for it, because customers could not get any food from the canteen without first paying for a meal coupon. Rather, by the union's account, an argument had broken out when the canteen manager refused to give the striking workers drinking water. In the midst of a heated verbal exchange between manager and strikers, one worker accidentally dropped a glass. Police on patrol nearby heard the sound of the glass shattering on the floor and, believing wrongly that the workers were trying to wreck the place, did a lathi-charge and then opened fire. The union representative alleged that stray police bullets entered the office of the camp manager and killed his assistant, who was not even on strike.[55]

These conflicting accounts of the police shooting were but the opening salvo of a propaganda war waged over the next month between project

box 19, ibid.; "Tārāpur anu sanyantra ke karmcāriyoṁ dwārā pathrāv: 4 Amrīkiyon sahit 16 ghāyal," *Hindustān*, December 29, 1965; "Americans did not fire on strikers," *HT*, January 6, 1966.

[54] "Tārāpur ke haṛtālī majdūroṁ kā updrav: pulis kī golībār meṁ 8 mare," *Hindustān*, December 31, 1965; "8 Workers Die in Police Firing," *HT*, December 31, 1965; "Eight Die in Police Firing in Tarapur," *TOI*, December 31, 1965.

[55] "Tārāpur meṁ shānti; Fernāṇḍej kisī ko bhī pañc mānne ko taiyār," *Hindustān*, January 1, 1966.

management, police, and their sympathizers on the one hand, and union leaders, socialists, and their supporters on the other. An unsigned editorial printed in the *Times of India* on January 3, 1966 came down firmly on the side of police and management by appealing to national solidarity in a time of crisis. "An enlightened public opinion should not only endorse the action taken by the police," the editorial argued, "but support the suppression of lawlessness which is at all times dangerous but more so when there is a national emergency."[56]

Meanwhile, George Fernandes continued to argue energetically for the workers' cause, revealing additional details and making further accusations in a series of press conferences. In his telling, the outrages perpetrated by management and police were boundless. Workers protesting steep prices at the canteen had been beaten up by management goons. Police had shot and killed a certain Joy Mathew when he was asking them not to shoot. The police had also been indiscriminate in their use of tear-gas shells, even firing them at Sikhs praying on Guru Govind Singh Jayanti. Not only the police but also the American contractors had wielded tear gas. (In one press conference, Fernandes displayed some used shells to reporters.) The Americans had also brandished revolvers and shot sickening gas (adamsite) canisters at the strikers. According to Fernandes, the same type of sickening gas was also being used by the US military in Vietnam.[57]

As was common in Bombay at the time, organized labor staged massive protests against the police, calling for a judicial inquiry into the shooting. Two days after the shooting, a large group of workers representing the city's various unions set out from Azad Maidan near Victoria Terminus in Bombay. The destination of the procession was the Sachivalaya (state government secretariat), where Fernandes wanted to deliver the workers' demands to the government. Only a few blocks after setting out, the workers were stopped by a police barricade near the University of Bombay. (The police did allow Fernandes to go through the barricade and deliver his demands.) The following week, another procession marched on the American consulate. Hind Mazdoor Panchayat secretary Jagannath Jadhav delivered a memorandum addressed to the American Consul

[56] "Stamp it out," editorial, *TOI*, January 3, 1966.

[57] "Tārāpur meṁ rivālvar prayukt karne kā amrīkiyoṁ par ārop," *Hindustān*, January 2, 1966; "Committee for Relief of Stranded Tarapur Workers," *TOI*, January 1, 1966.

General, then made a speech to the assembled workers from the roof of a station wagon parked in front of the consulate.[58]

Shortly after the police shooting, Fernandes began calling for mediation in the management-labor dispute by a neutral third party such as the Swatantra Party, Sarvodaya Mandal, or the American Peace Corps. The Bombay Sarvodaya Mandal, a Gandhian organization, was one of several groups that went on fact-finding missions to Tarapur. From its investigations, the Sarvodaya Mandal found that both the police and the Bombay Labour Union had made baseless accusations: strikers had not tried to raid the explosives magazine and Americans had not fired any sickening-gas shells at strikers. Fernandes apologized for having accused the Americans of using sickening gas, but he continued to condemn the police for their use of tear gas against their fellow countrymen. Separate investigations by the Praja Socialist Party found that the tear-gas canisters used in the confrontation were up to five years past their expiration date.[59]

Throughout the month of January 1966, opposition members in government called for mediation of the dispute. At the encouragement of Swatantra Party secretary Madhu Mehta, George Fernandes finally met with R.L. Portlock, Bechtel Project Manager, in Bombay on Saturday, January 29. This was all it took to break the impasse. Bechtel agreed to recognize Bombay Labour Union, and in response, the union called off the strike and work resumed on Monday.[60]

Despite achieving recognition, the outcome was at most a half-victory for the Bombay Labour Union. As part of the compromise that ended the strike, the union agreed that it would stage no further strikes, thus hobbling its ability to apply pressure on management. Also, while Bombay Labour Union was recognized, management's arrangement with the INTUC-affiliated Tarapur Anushakti Kamgar Sangh also remained in

[58] "Protest Morcha in City Today: Tarapur Firing," *TOI*, December 31, 1965; "Direct Action; Threatened: Bombay Unions," *TOI*, January 1, 1966; "Workers March to Consulate: SSP-HMP Morcha," *TOI*, January 6, 1966.

[59] "Workers March to Consulate: SSP-HMP Morcha," *TOI*, January 6, 1966; "PSP Urges Judicial Probe into Tarapur Firing," *TOI*, January 10, 1966.

[60] TAPP quarterly report 6, January–March 1966, box 20, USAID Closed Project Loan Files 1960–1971, RG 286, NARA; "'Union Must Call off Strike': Step for Tarapur Normalcy," *TOI*, January 10, 1966; "Tarapur Strike Is Called Off," *TOI*, January 30, 1966.

effect. After further negotiation, management signed an agreement with both unions on June 3, 1966, but about fifty workers declined to join either union.[61]

Radical organized labor at Tarapur Atomic Power Project had proven to be weaker than it at first appeared. Bombay Labour Union could bring the Bechtel workforce to a standstill for two months and organize massive demonstrations and strikes in Bombay, but it was unable to win an exclusive collective bargaining agreement with project management. The radicals' demonstrations turned out to be more theatrical performance than exercise of real power.

As a coda to the story of the Tarapur strike, it is worth noting that radical labor continued to stage its enormous strikes and demonstrations into the 1970s, with George Fernandes always at the forefront. In May 1974—the same month as the Peaceful Nuclear Explosion—one million railway workers walked off the job for three weeks, crippling the national transportation network. It was the biggest strike of Fernandes's career. After serving a prison term in the Emergency for plotting to overthrow Indira Gandhi's dictatorship, Fernandes served in the short-lived Janata coalition government of 1977–80. Years later, his political career took a right turn when he served as Defence Minister for the Bharatiya Janata Party government of Atal Bihari Vajpayee. His radical socialist past long behind him, George Fernandes died in 2019 at the age of 88.[62]

Tarapur's Troubles

When construction of Tarapur Atomic Power Station began at the end of the 1964 monsoon, the project timetable anticipated completion and commissioning of the station in October 1968. The 1965 Indo-Pakistan War and the Bombay Labour Union strike could have made the project slip behind schedule, but neither did. Nevertheless, other problems relating to personnel as well as hardware caused the project to be

[61] Tarapur Atomic Power Project Meeting Minutes, July 13, 1966, box 24, USAID Closed Project Loan Files 1960–1971, RG 286, NARA.

[62] Ramachandra Guha, *India After Gandhi: The History of the World's Largest Democracy* (New York: Ecco, 2007), 479, 503–4, 534, 664; "Ex-union Minister George Fernandes Dead at 88," *Bihar Times*, January 30, 2019.

commissioned a full year behind schedule. Then for years after its completion, Tarapur was beset by a string of technical problems that forced it to shut down again and again.

In July 1966, project welders, frustrated with the pace of work required to catch up after the Bombay Labour Union strike, staged a wildcat strike unauthorized by either of the unions recognized by management. Bechtel promptly fired the strikers, but a shortage of qualified welders in India ultimately forced management to agree to rehire the welders in September. The welders' strike had none of the dramatic episodes of the earlier strike—no violent encounters, no solidarity marches by sympathetic labor groups in Bombay—but it had a bigger impact on the project because it made construction begin to slip behind schedule.[63]

As the plant took shape, technical problems too began to hamper its completion and commissioning. In late 1967, after cracks appeared in the reactor vessel of a similar GE plant in New Jersey, Oyster Creek, dye-penetrant tests revealed cracking in both units at Tarapur. Dismantling the reactor cores and grinding down and rewelding the affected parts further delayed the project.[64]

At 8:37 on the morning of February 1, 1969, Reactor 1 of Tarapur Atomic Power station went critical for the first time. Reactor 2 followed at the end of the month, and the station began generating power on a trial basis on the first of April. It took several months until TAPS was ready to go into full commercial operation. Plant operators put it through its paces in a series of tests, culminating in a hundred-hour trial. The Maharashtra State Electricity Board had just completed an on-site switchyard, which linked TAPS to the Maharashtra and Gujarat state grids. In May, and again in August, the switchyard malfunctioned, interrupting electric supply and causing the station to shut down. The problems turned out

[63] TAPP quarterly report 8, July–December 1966, box 20, USAID Closed Project Loan Files 1960–1971, RG 286, NARA; TR, July 22, 1966; TR, September 9, 1966.

[64] TAPP quarterly report 13, October–December 1967, box 20, USAID Closed Project Loan Files 1960–1971, RG 286, NARA; D.C. Bertossa, J.E. Corr, and W.L. Walker, "Stress Corrosion Cracking of Stainless Steel Components, Tarapur Atomic Power Plant," APED-5627, box 24, ibid.; "Cracks in Equipment Delay Starting of Tarapur Unit," *TOI*, October 1, 1968.

to have been caused by faulty insulators that led to short-circuiting in the wet monsoon weather.[65]

Then came an economic problem. When the station was almost ready for commercial operation, the Department of Atomic Energy planned to run Reactor 1 at full capacity and divide evenly the 200 MW produced between Gujarat and Maharashtra. Gujarat was a willing customer of its full allotment of 100 MW, but Maharashtra State Electricity Board reported that it could use no more than 25 MW. The monsoon rains had been good and the reservoirs were full, so the state wanted to use hydropower rather than power from Tarapur. "What are they doing to Tarapur? Trying to wreck the whole concept of nuclear power?" cried an editorial in the *Times of India*. The editorial said that Maharashtra's rejection of Tarapur power was "scandalous on its own count because the monsoon was not exactly a surprise and the State had always cried out for more power. The rains didn't come and suddenly fill the hydel reservoirs to an unexpectedly high level."[66]

General Electric finally turned the plant over to the Government of India on October 28, 1969. Prime Minister Indira Gandhi formally dedicated the plant the following January, claiming in her speech that the plant's construction represented the partial fulfillment of the dreams of two great men, her father and Homi Bhabha.[67] In more ways than one, the Prime Minister was correct in this assertion. In the construction of Tarapur Atomic Power Station, India's technocrats had brought one of the highest-of-high technologies to their country. To them, Tarapur represented the triumph of modernity over tradition, backwardness, and economic underdevelopment. Yet Tarapur also represented the hollow promises of Nehruvian industrialization. Rural people had been displaced for the construction of a power station that would benefit the urban elite. The construction of the station had deepened India's foreign debt. And

[65] TAPP quarterly report 18, January–March 1969, box 20, USAID Closed Project Loan Files 1960–1971, RG 286, NARA; TAPP quarterly report 19, April–June 1969, ibid.; "Tarapur Reactors Shut Down: Switchyard Defect," *TOI*, August 11, 1969.

[66] "State Unable to Take Full Quota of Tarapur Power," *TOI*, August 2, 1969; "Sheer Bungling," editorial, *TOI*, August 13, 1969.

[67] Indira Gandhi, "Tarapore: A Dream Come True," in *Selected Speeches of Indira Gandhi, vol. 2: The Years of Endeavour, August 1969–August 1972* (New Delhi: Publications Division, Ministry of Information and Broadcasting, Govt. of India, 1973), 421–24.

the station never worked as well as its advocates had claimed it would. For years after its commissioning, TAPS suffered from a series of hardware failures and safety problems that left it under a cloud of public suspicion and mistrust. A worker died in an accident in a nitrogen chamber just six months after the dedication. During the first refueling of the station in 1971 and 1972, both units were shut down for long periods for repairs and modifications—eight months for Reactor 1 and six for Reactor 2. With reduced supply from Tarapur, the Maharashtra State Electricity Board had to reduce its transmission voltage and frequency, causing electric clocks in Bombay to run slow.[68]

The Department of Atomic Energy defended the problems as mere teething troubles inherent in a project as big as Tarapur, but a growing body of detractors of atomic energy in India began to suspect that these were issues inherent to the plant and nuclear technology overall. Other opponents, in India as well as the United States, came to believe that the Indian government was incapable of managing nuclear plants safely. In February 1976, the inaugural issue of the American magazine *Mother Jones* published a devastating (if incredible) exposé of safety problems at the plant, including allegations of large numbers of radiation deaths and an extensive government cover-up. Responses to the article fell on predictable lines. Members of the US Congress asked the Nuclear Regulatory Commission to delay shipments of enriched uranium to India, while the Indian Atomic Energy Commission claimed that the accusations were nothing but lies.[69]

[68] "One Dies in Tarapur Station Mishap," *TOI*, July 30, 1970; "Clocks Will Go Slow for 3 Weeks: Low Voltage," *TOI*, May 1, 1971; K.C. Khanna, "Atomic Power Prospects I—The Importance of Tarapur," *TOI*, June 23, 1972.

[69] M.V. Kamath, "Doubts about Tarapur Baseless: Sethna," *TOI*, February 20, 1976; K.C. Khanna, "Tarapur's Safety Record: AEC Rebuts False Charges," *TOI*, March 15, 1976.

Conclusion

As the first nuclear plant built in India, Tarapur Atomic Power Station attracted more attention than contemporary conventional power projects. Newspapers across the country printed updates of the plant's construction, commissioning, and interminable technical problems. Papers in London reported on the Bombay Labour Union strike, while American papers carried stories about the plant's financing. Film crews from Indian News Review and the United States Information Service made documentaries about the construction project.[70]

When it was built, Tarapur was a symbol of the new and the modern. A General Electric advertisement that appeared in Indian magazines shortly after the company's selection as prime contractor characterized the construction of Tarapur Atomic Power Station as: "A dreaming old-world village wakes upto [*sic*] atomic age!"[71] Over time, though, the novelty of nuclear power wore off, and the vanguard of modernity shifted from industrial megaprojects to information technology and renewable energy. After TAPS Reactors 1 and 2, India has completed twenty-one more reactors (as of early 2021) at a total of seven sites, including two additional reactors at Tarapur.[72] Yet seven nuclear plants are a pittance compared with the scores of hydroelectric projects completed since independence (as discussed in Chapter 7). Even with environmental exclusion zones around the reactor buildings (inside which no people are allowed to live), nuclear plants do not take up as much land as hydroelectric projects, with their farmland- and village-obliterating reservoirs. Although a nuclear accident could have catastrophic results—as seen in 1986 at Chernobyl in the Soviet Union—the Indian nuclear plants have operated

[70] TR, July 23, 1965; TR, June 17, 1966; USIS, "Nuclear Power from Tarapur," 1969, US Moving Images compiled 1982–1999, RG 306, NARA; "Eight Killed in Strike Riots at Atomic Plant," *The Times* (London), December 31, 1965; "GE Signs India Contract," *Wall Street Journal*, May 11, 1964.

[71] The advertisement was placed by International General Electric and appeared in the October 1963 issue of the *Indian Journal of Power and River Valley Development*.

[72] Power Reactor Information System, "India," *International Atomic Energy Agency*, https://pris.iaea.org/PRIS/CountryStatistics/CountryDetails.aspx?current=IN (accessed February 26, 2021).

without major incident and have attracted less organized public opposition than the large dams.[73] Nuclear plants have thus not affected the Indian national consciousness in the same way as hydro plants.

Nuclear power also never became the technological boon that India's planners had anticipated. India's twenty-three commercial power reactors have played only a small role in the country's development. In January 2021, nuclear power represented a mere 1.8 percent of installed capacity in the country, compared with 12.2 percent for hydroelectricity and 61.5 percent for fossil fuels.[74]

India was eventually able to manufacture nuclear reactors domestically and develop a domestic source of natural-uranium fuel. So what happened to the promise of nuclear power in India? Despite the domestic reactors and fuel, nuclear power remains a bad technological fit for India, compared with hydroelectricity. Even if most of the components used in a plant are produced domestically, nuclear power remains far more capital-intensive than hydroelectricity. As discussed in Chapter 7, India's seasonal monsoon rains make generation of hydroelectricity from reservoir storage practical in most of the country. Dams can be built mostly from domestic materials, and India's huge labor force could be used in the early independence period in place of many types of machines, thereby saving foreign exchange.

By comparison, there are fewer good sites for nuclear plants in India, and nuclear power development has been concentrated in certain areas. With the exception of Narora Atomic Power Station in Uttar Pradesh and Rajasthan Atomic Power Station, India's nuclear plants have, like Tarapur, been built on or near the coast, in an arc from Tamil Nadu to Gujarat. There are plans for nuclear plants in other states, including Madhya Pradesh and West Bengal, but it remains to be seen when, if ever, these plants will be built.

[73] There have been some anti-nuclear demonstrations in India. See, for example, Praful Bidwai, "People vs Nuclear Power in Jaitapur, Maharashtra," *Economic and Political Weekly*, February 19, 2011, 10–14.

[74] "Power Sector at a Glance, All India," *Government of India Ministry of Power*, https://powermin.nic.in/en/content/power-sector-glance-all-india (accessed February 26, 2021).

References

1. Abraham, P. "Safety Features at Tarapur." *The Indian and Eastern Engineer*, (February 1970), 117–20.
2. Allen, Charles, and Michael Mason. *Plain Tales from the Raj: Images of British India in the Twentieth Century*. New York: St. Martin's Press, 1976.
3. Bayley, David H. "The Pedagogy of Democracy: Coercive Public Protest in India." *The American Political Science Review* 56, no. 3 (September 1962): 663–72.
4. Bayley, David H. "The Police and Political Order in India." *Asian Survey* 23, no. 4 (April 1983): 484–96.
5. Chakravarti, M.N. "Tarapur Atomic Power Project." *IJPRVD* 18, no. 2 (February 1968): 41–47.
6. Cohen, Benjamin B. *In the Club: Associational Life in Colonial South Asia*. Manchester: Manchester University Press, 2015.
7. Datta Chaudhuri, Mrinal. "Labor Markets as Social Institutions in India." *Iris-India Working Paper* 10 (March 1996). http://pdf.usaid.gov/pdf/Pnaby353.pdf.
8. Finnie, Richard. *Bechtel in Arab Lands: A Fifteenth-Year Review of Engineering and Construction Projects*. San Francisco: Bechtel, 1958.
9. Gandhi, Indira. "Tarapore: A Dream Come True." In *Selected Speeches of Indira Gandhi, vol. 2: The Years of Endeavour, August 1969–August 1972*, 421–24. New Delhi: Publications Division, Ministry of Information and Broadcasting, Govt. of India, 1973.
10. Ganguly, A.K. "U.S. Engineer Helps Develop India's Power Resources." *Indian Railway Gazette* (July 1963): 287–88.
11. Guha, Ramachandra. *India After Gandhi: The History of the World's Largest Democracy*. New York: Ecco, 2007.
12. Hecht, Gabrielle. *The Radiance of France: Nuclear Power and National Identity after World War II*. Cambridge, MA: MIT Press, 1998.
13. Ingram, Robert L. *The Bechtel Story: Seventy Years of Accomplishment in Engineering and Construction*. San Francisco: Bechtel, 1968.
14. James, Ralph. "Politics and Trade Unions in India." *Far Eastern Survey* 27, no. 3 (March 1958): 41–45.
15. Kalhan, Anil, Gerald P. Conroy, Mamta Kaushal, Sam Scott Miller, and Jed S. Rakoff. "Colonial Continuities: Human Rights, Terrorism, and Security Laws in India." *Columbia Journal of Asian Law* 20 (2006): 93–234.
16. Mathur, Kuldeep. "The State and the Use of Coercive Power in India." *Asian Survey* 32, no. 4 (1992): 337–49.
17. Morgan, W., and A.G. MacPherson. *The Indian Penal Code (Act XLV of 1860) with Notes*. Calcutta: G.C. Hay, 1863.
18. Morris, Morris David. "Labor Discipline, Trade-Unions, and the State in India." *Journal of Political Economy* 63, no. 4 (August 1955): 293–308.

19. Ramachandran, V., et al. “Design and Construction of the 120 Metre High Meteorological Tower Facility at the Tarapur Atomic Power Station.” *The Journal of the Institution of Engineers (India)* 48, no. 9 (September 1968): 268–80.
20. Rotter, Andrew J. *Comrades at Odds: The United States and India, 1947–1964*. Ithaca, NY: Cornell University Press, 2000.
21. “Tarapur: Gateway to the Nuclear Age.” *Participant Journal* (January 1970): 12–14.

CHAPTER 10

Conclusion

On May 27, 1964, Prime Minister Jawaharlal Nehru died in his bed at Teen Murti Bhawan in New Delhi. Still several months short of his seventy-fifth birthday, Nehru had been a haggard, exhausted man when he collapsed from a heart attack on the morning of May 27.[1] He had lived to see his dream of nonaligned India as the moral leader of Asia shatter against the military threat of China.

Yet another of Nehru's great dreams—the dream of a prosperous, egalitarian India marching toward the future under the guidance of a socialist command economy—was still very much alive. Granted, Nehruvian socialism had already begun to show some weaknesses, in its over-reliance on foreign financing and food aid since the late 1950s. And there was the more fundamental problem of how development dominated by an urban elite could be accountable to India's predominantly rural population. But at least in the eyes of the Congress establishment, the socialist command economy still offered the only way forward for India.

Nehru was succeeded as Prime Minister by Lal Bahadur Shastri, every bit the Nehruvian in political outlook. As Prime Minister, Shastri continued to promote the Nehruvian agenda. Yet he only held office for less than two years, before succumbing to a heart attack while in the

[1] Ganesh Kudaisya, "Death of a Democrat," *History Today*, May 21–27, 2014.

W. A. T. Logan, *A Technological History of Cold-War India, 1947–1969*, Palgrave Studies in the History of Science and Technology,
https://doi.org/10.1007/978-3-030-78767-7_10

Soviet Union during peace talks with Pakistan after the 1965 war. In looking for a successor to Shastri, Congress leadership picked Nehru's daughter Indira Gandhi, thinking wrongly that she would be pliable and easy to manipulate.

The Third Five-Year Plan ended the same year that Indira Gandhi became Prime Minister. The Planning Commission had hoped that the Third Plan would be the first step toward India's attaining economic and technological autarky, but a spate of crises had disrupted the plan. The wars with China in 1962 and Pakistan in 1965 had led to accelerated military spending, which drained resources away from civilian projects. There was also a massive drought in 1965 and 1966, which led to crop failures and the threat of famine. In response to these crises, and to promote Indian exports, the Government of India devalued the rupee in 1966 from 4.76 to 7.50 to the US dollar.[2]

The Fourth Five-Year Plan was originally set to start in 1966, but in response to the crises, the Planning Commission shelved the plan for three years. A revised plan, which rolled out in 1969, called for a total public-sector outlay of 144 billion rupees ($19.2 billion), compared to 75 billion rupees for the Third Plan, but in absolute terms, this only represented a modest 22 percent increase over the previous plan because of the rupee's devaluation. Superficially, the Fourth Plan resembled the Third, with resource allocations for most categories differing from the earlier plan by only one or two percentage points. But at its core, the Fourth Plan was less ambitious, with more emphasis on agriculture than heavy industry.[3]

What had changed in the two decades since independence? To begin with, India's large population had continued to grow. The 1971 census would record 547,949,809 people in India, an increase of nearly 75 percent since independence. The economy and industry had grown as well. The national income in 1967–1968 was 173 billion rupees. India now boasted 64 million acres of farmland under irrigation, 10,189 MW

[2] Dharma Kumar, ed., *The Cambridge Economic History of India* (Cambridge: Cambridge University Press, 1983), 2:956; Nick Cullather, *The Hungry World: America's Cold War Battle Against Poverty in Asia* (Cambridge, MA: Harvard University Press, 2010), 205, 231; Rahul Mukherji, "India's Aborted Liberalization—1966," *Pacific Affairs* 73, no. 3 (Autumn 2000): 377.

[3] *Cambridge Economic History of India*, 2:956; Satish C. Seth, "'Public Sector and Co-operation': As Seen in the Draft of the Indian Fourth Five-Year Plan (1968–74)," *Annals of Public & Co-operative Economy* 40, no. 4 (October 1969): 487–501; Sham Lal, ed., *The Times of India Directory and Yearbook Including Who's Who 1969* (Bombay: Times of India Press, 1969), 260.

of electrical generative capacity, and 36,870 route-miles of railways. The industrial labor force had grown to 4.6 million, an increase of nearly three times since 1931. India's economy would never be the autarkical bastion that the Nehruvians envisioned, but import-substitution programs for metals, chemicals, and machinery were beginning to have some success. Indian industry in 1967 produced, among other goods, 1.7 million bicycles, 457,000 sewing machines, 670,000 tons of paper, and 12 million tons of cement.[4]

Overall economic growth notwithstanding, the 1960s had been a decade of crises, and there were more crises and changes to come in the 1970s and 1980s. In 1971, Pakistan responded to unrest in its eastern wing with a brutal, bloody crackdown that killed hundreds of thousands of Bengalis. Later that year, Indira Gandhi, having consolidated her power domestically, waged a war on Pakistan that led to the dismemberment of the country and the birth of Bangladesh as an independent nation. Throughout the war, the administration of US President Richard Nixon had frozen aid to India and unequivocally supported Pakistan, genocide and all. In response to a rising tide of anti-American public sentiment in India, the Ministry of External Affairs served notice to USAID in May 1972 that all American aid workers had to leave the country within a month. The actual withdrawal took a little longer, but the last American technical experts were gone by mid-1973, thus bringing the age of US technical assistance to India to an end.[5]

Despite her popular political accomplishments early in the decade, by the mid-seventies Indira Gandhi was feeling threatened by domestic opposition, so she declared an Emergency and ruled for a year and a half as dictator. At the end of the Emergency in 1977, Congress was decisively voted out of power at the national level for the first time since independence. The party that bested Congress was a coalition known as the Janata Party. Consisting of a diverse cast of anti-Congress (or at any rate, anti-Indira) figures on the right and left, Janata proved to be an unstable coalition. In 1980, it was swept back out of power by a resurgent Gandhi. The early eighties witnessed unprecedented ethnic violence in Assam and a Sikh insurgency in Punjab, the latter of which the army suppressed with

[4] Lal, *TOI Yearbook 1969*, 76, 85, 90, 117, 236, 255–58; *Cambridge Economic History of India*, 2:973; Sham Lal, ed., *The Times of India Directory and Yearbook Including Who's Who 1973* (Bombay: Times of India Press, 1973), 143.

[5] David C. Engerman, *The Price of Aid: The Economic Cold War in India* (Cambridge, MA: Harvard University Press, 2018), 331–34.

force. The army's violent intervention in the Golden Temple in Amritsar led to Gandhi's assassination by her own bodyguards in 1984.[6]

During the decades of crisis, the Indian government maintained an ideological commitment to autarky, but the reality was, as usual, more complicated than the ideal. In 1978, the Janata government forced American firms IBM and Coca-Cola out of India by making them subject to the 1973 Foreign Exchange Regulation Act, which would have made it prohibitively expensive for them to continue to operate in the country. Other foreign businesses, though, were spared from the act and allowed to stay.[7]

One of these was Union Carbide India Ltd. (UCIL), which had operated in India since 1934. The Janata government allowed UCIL's American parent company to maintain 51-percent ownership of its subsidiary, on the grounds that the company's operations were providing technology transfer into India. At this time, Union Carbide was building a pesticide plant in the new industrial center of Bhopal, to supply India's green-revolution agricultural sector. The plant had been operating for only a few years when, on the night of December 2–3, 1984, an operational error led to the release of toxic methyl isocyanate into the atmosphere. The poison gas drifted for miles across the city, causing residents to cough, vomit, or lose consciousness and die. Survivors fled their homes in the middle of the night. All told, the gas killed thousands of residents of Bhopal and caused permanent health damage for many thousands more. It was the deadliest industrial disaster in history.[8]

[6] For the violence in Assam, see Shekhar Gupta, *Assam, a Valley Divided* (New Delhi: Vikas, 1984).

[7] Ramachandra Guha, *India after Gandhi: The History of the World's Largest Democracy* (New York: HarperCollins, 2007), 387, 523, 683; Raju Chellam, "Where Computers Dare," *Times of India* (*TOI*), November 16, 1986.

[8] B.R. Tomlinson, "Foreign Private Investment in India 1920–1950," *Modern Asian Studies* 12, no. 4 (1978): 655–77; "U.S. Credit for Bhopal Pesticides Plant," *TOI*, January 11, 1976; "375 Die in Bhopal Gas Leak," *TOI*, December 4, 1984; "Low Farce," *Economic and Political Weekly*, December 15, 1984, 2097.

The Technological Heritage of Nehruvian India

The dams, canals, factories, rail lines, and other pieces of infrastructure and technological artifacts of the Nehruvian era continued to be used for decades after Nehru's death. Many of them are still in use to this day, although some have been decommissioned and left to go to rust. As of early 2021, the projects still in use include Saraighat Bridge, Umiam Dam, and Tarapur Atomic Power Station.

For nearly six decades, Saraighat Bridge has served as a vital transportation link in Assam, but its very uniqueness and indispensability make it a scene of daily traffic jams and a strategically vulnerable structure. While army chiefs have recommended boring tunnels under the Brahmaputra rather than building bridges across it, the railways and highway department have built several other bridges across the Brahmaputra in Assam as well, including a second span immediately downstream of the original Saraighat Bridge. Completed in 2016 by the contractor Gammon-India, the second Saraighat Bridge has a single road deck crossing reinforced concrete arch spans. The new bridge carries two lanes of northbound traffic across the river, while both lanes on the road deck of the old span now carry southbound traffic. (The rail deck has, since 1984, had a single broad-gauge rail line rather than the dual meter-gauge lines originally installed when the bridge was built.)[9]

In the hills above the Brahmaputra Valley, Umiam Dam holds an ambivalent place in the life of Shillong and the state of Meghalaya. The power project is a critical source of electricity for the state. The reservoir is a place for recreation, and reservoir levels are printed daily in the *Shillong Times*. Yet the dam and reservoir are also a cause of concern—like large dams across India. The lake is becoming increasingly polluted by runoff from Shillong. The *Shillong Times* has raised the alarm about siltation in the lake—visible when shoals appear during the dry season—or mechanical problems in the powerhouse's aging machinery. In the 2010s, the state utility launched a major refurbishment program for Umiam

[9] Rahul Karmakar, "Build Tunnels Under Brahmaputra River to Outfox Enemy Forces, Says Army," *HT*, May 23, 2017; "Only Second Bridge Can Ease Traffic on Saraighat," *Assam Tribune*, July 14, 2015; "List of Current Projects Having Contract Value More Than 100cr.," *Gammon India*, http://www.gammonindia.com/projects/current-project-for-gammon-india.htm (accessed November 9, 2015); Samudra Gupta Kashyap, "Nitin Gadkari Inaugurates New Brahmaputra Bridge, Sets Rs. 65,000 Crore for New Projects in 5 Years," *Indian Express* (*IE*), January 30, 2017.

Stage I, in collaboration with Japanese consultants. Japanese industry had provided the original turbine-generators in the 1960s, and a half-century later Japanese expertise was invited back to extend the life of the project.[10]

Since the inauguration of Umiam Stage I in 1965, Indian industry has developed the capability to produce most of the machinery and specialized equipment that had to be imported for the first stage. Public-sector Bharat Heavy Electricals Ltd. manufactures turbine generators, transformers, switchgear, and other equipment needed for hydroelectric plants. Bharat Earth Movers Ltd., another public-sector firm, makes heavy machinery for construction projects. Private-sector foreign companies such as JCB, Caterpillar, and Hyundai also produce construction equipment at factories in India. For the third and fourth stages of Umiam Hydroelectric Project, completed in 1979 and 1992 respectively, the Meghalaya State Electricity Board did not need to import turbines from abroad; it could buy Indian.[11]

Like Umiam Dam, Tarapur Atomic Power Station has also been a target of controversy and concern. As the first nuclear plant built in a nonaligned country, Tarapur was controversial from the beginning. The original fuel supply agreement with the United States met opposition in both countries. American opponents objected to the prospect of sending nuclear material to a nonaligned nation, and they insisted on strict safeguards for monitoring and inspecting spent fuel to ensure that it was not diverted to weapons production. Indian opponents, for their part, objected to the inequality inherent in the safeguards.

ThePeaceful Nuclear Explosion in May 1974 did not lead to the immediate end of American nuclear assistance to India (as it did for Canadian assistance), but it did contribute to a climate of mutual suspicion and mistrust between India and the United States. India's nuclear test raised the prospect of dozens of other countries' gaining access to nuclear weapons as well. According to US Congressman Clarence D. Long (Democrat, Maryland), this would upset the cold-war balance

[10] Paramita Sarma, Letter to the Editor, *Shillong Times*, March 17, 2014; "Leakage in Umiam Power House," *Shillong Times*, November 1, 2011; "Govt Looking into Water Leak at Umiam, Says Mondal," *Shillong Times*, November 9, 2011; Eladbor Laloo, "What Ails the Umiam Reservoir?" *Shillong Times*, November 27, 2012; Amitangshu Acharya, "Umiam, Shillong's 'Wasted' Pride," *Calcutta Telegraph*, September 30, 2009; Central Board for the Prevention and Control of Water Pollution, *Environmental Status: Barapani Lake, Meghalaya* (New Delhi, n.d. [1982–83]).

[11] "Existing Power Stations," Meghalaya Energy Corporation Limited, http://meseb.nic.in/network.htm (accessed April 18, 2012).

of power and could lead to a complex, dangerous, and unmanageable international situation. In 1978, the US Congress passed the Nuclear Non-Proliferation Act, which mandated that the United States cut off nuclear assistance to any country that did not submit to IAEA inspections at all of its nuclear facilities, because spent fuel from a civilian reactor could be reprocessed into nuclear warhead material. This act specifically targeted India, because the Canada-India Reactor had no safeguards. In 1981, the United States unloaded its fuel supply duties for Tarapur on France, which had recently opened a large uranium enrichment plant dubbed "Eurodif." France agreed to maintain safeguards for the fuel at Tarapur.[12]

Tarapur was also a perennial subject of concern for safety and environmental reasons, as any radiation leak from the reactors could be an even bigger disaster than the Union Carbide chemical spill in Bhopal. After the *Mother Jones* exposé in 1976, Tarapur again emerged in the public spotlight in 2011. In March of that year, a Japanese nuclear plant named Fukushima Daiichi melted down when the Japanese coast was struck by a major tsunami. Fukushima was a GE plant built to a similar design as Tarapur, and it stood to reason that Tarapur was also vulnerable on its coastal site. Anti-nuclear journalist Praful Bidwai wrote an article for *Outlook* magazine that characterized the technocrats in charge of the Nuclear Power Corporation of India Limited (NPCIL, the operator of Tarapur since 1987) as "men who are totally cut off from reality and have a default conviction in their own omniscience and infallibility." He also criticized the GE-designed equipment that had failed at Fukushima and could fail at Tarapur as well. "Nuclear power has subjugated our energy policy and budgets to an unaccountable, self-perpetuating, pampered technocracy," Bidwai concluded. "The juggernaut must be halted."[13]

In a rejoinder to Bidwai's article, NPCIL claimed that Tarapur had an excellent safety record and had received international acclaim for its operations. According to NPCIL, "A large number of Parliamentary delegates and nuclear experts of Nuclear states have visited our nuclear

[12] Clarence D. Long, "Nuclear Proliferation: Can Congress Act in Time?" *International Security* 1, no. 4 (Spring 1977): 52–76; "Nuclear Non-proliferation Act of 1978, Public Law 95–242," *The American Journal of International Law* 72, no. 3 (July 1978): 712–42; Gursharan S. Dhanjal, *Tarapur: The Politics of Nuclear Age* (Delhi: Rajdhani Book Service, 1992), 18, 84–85, 88.

[13] Praful Bidwai, "India's Nuclear Neros," *Outlook*, March 28, 2011, https://www.outlookindia.com/magazine/story/indias-nuclear-neros/270948 (accessed February 24, 2020).

power station at Tarapur on various occasions and are impressed with the safety and overall performance of our reactors." Propaganda aside, NPCIL undertook an upgrade and refurbishing program to make Tarapur safer and protect against a parallel disaster to Fukushima. Despite calls to shut the plant down—and despite the fact that it was originally built for a planned lifespan of only 25 years—Tarapur Atomic Power Station reached 50 years of operation in 2019.[14]

Technology Transfer and Indigenization in Modern India

This book has been a story of technology transfer, indigenization, and use. In the case studies, I have tried to understand how technologies that originate outside of India can become Indianized. This could mean that the technology is produced in India, or it could mean simply that the technology takes on distinctively Indian uses and meanings. For domestic manufacture, full technology transfer—that is, transfer not just of technological artifacts but also the knowledge of how to make and modify the things in question—depends on several factors. One of them is having a big enough pool of engineers, designers, and other practitioners who have the requisite technical knowledge to create as well as operate. Another is the existence of an industrial base that can support domestic production of machinery and parts big and small for an indigenization or import-substitution program. A third factor is having capital to set up domestic production—which may or may not prove to be cheaper than simply importing something in the short or even the long term.

India's experience with domestic arms manufacture illustrates the complexities and challenges inherent in indigenization or import-substitution. For the Indian Air Force in the Nehruvian era, the most ambitious import-substitution project was the HF-24 Marut, the indigenous fighter jet. The HF-24 failed to live up to expectations because the engine for which it was originally designed, the Orpheus 12, never entered production, and Hindustan Aeronautics had to make do with the lower-performance Orpheus 703 instead. Indian industry could not at once produce an airframe, a new high-performance engine, and enough

[14] "Our Comments on Your Story 'India's Nuclear Neros,"' *NPCIL*, http://npcil.nic.in/pdf/news_25mar2011_02.pdf (accessed January 11, 2016).

parts of each to make a domestic jet any cheaper than an imported model. Only countries with large, well-developed military-industrial complexes like the United States or Soviet Union could afford to make substantially indigenous jets.[15]

The IAF retired its HF-24 squadrons in the early eighties, but the indigenous fighter dream did not disappear. Its most recent incarnation is the LCA (Light Combat Aircraft) or "Tejas" (Radiant), envisioned as a domestic replacement for the IAF's aging MiG-21 fighters. The LCA program began in 1983, but it was not until 2001 that a prototype made its first flight, and it would take another fifteen years before the first two production airplanes finally entered squadron service. In a near-repeat of the IAF's turning to Sukhoi fighters in the late 1960s, delays in the LCA program compelled the IAF to order thirty-six French-made Dassault Rafale fighters in the interim.[16]

The Dassault fighters are representative of a trend in Indian military history: importing. Far from attaining self-sufficiency in arms production, India has become one of the world's largest arms importers, with the majority of its arms originating in Russia. According to the Stockholm International Peace Research Institute (SIPRI), India accounted for 14 percent of global arms imports in the years 2009–2013, the largest amount for any country. China, with 5 percent of arms imports, came in as a distant second. (In 2019, SIPRI reported that Saudi Arabia had surpassed India as the world's largest arms importer.) India is a large country with ambitions of being the dominant power in its region, while rivals Pakistan and China have vested interests in countering Indian dominance. Holding these ambitions, but lacking the industrial capacity to manufacture many of its own weapons, India has had to look abroad to fulfill its weapons needs since the early independence period.[17]

[15] "India's Supersonic Aircraft Registers 30% Improvement," *India News*, October 14, 1966; "Replacement of Aircraft," Oral Answers, *Lok Sabha Debates* (*LSD*), September 16, 1981, 20–21.

[16] "India, US Finalise Rs. 3000 cr Jet Engine Deal," *HT*, January 20, 2013; Chethan Kumar, "32 Years on, HAL Finally Hands over Tejas to IAF; Induction Still Far Away," *TOI*, January 18, 2015; "Agreement on Rafale Price Will Take Another Month's Bargaining," *TOI*, January 26, 2016; "Mig 21 kī jagah lenge ghātak Tejas vimān," *Rājasthān Patrikā*, February 2, 2016; Pramit Pal Chaudhuri, "Tejas Deserves One Cheer Today, Maybe Two Later," *HT*, July 1, 2016.

[17] Stockholm International Peace Research Institute, "International Arms Transfers," *SIPRI Yearbook 2014*, https://sipri.org/yearbook/2014/05; Gardiner Harris, "World's

As mentioned earlier, some Indian import-substitution programs have been more successful than arms manufacture, including metals, chemicals, and machinery. Indian public and private industry now produces steel, cast iron, pharmaceuticals, automobiles, motorcycles and motor-scooters, locomotives, subway trains, and heavy electrical equipment, among an enormous array of other products. These industries have succeeded where others have failed because all three factors have been in place to support them: knowledge (sometimes brought in by technology transfer agreements with foreign companies), a base of supporting industries, and capital (also sometimes foreign in origin).

Nehruvian Development in Retrospect

Twenty-first century India is not Nehruvian India. Jawaharlal Nehru has been dead for more than fifty years, and his dream of a socialist India is dead as well. Nehru's grandson Rajiv Gandhi, who succeeded his mother Indira as Prime Minister after her assassination, started to roll back the socialist command economy in the mid-eighties. In 1991, Manmohan Singh, Finance Minister for P.V. Narasimha Rao, completed the dismantling of the command economy in response to a financial crisis.[18] Economic liberalization, as it was called, opened the door to greatly expanded private-sector investment, especially by foreign firms. Soon American fast-food chain restaurants opened in major Indian cities, and German cars began to ply the Indian roads.

Since 1991, and especially since the election of Narendra Modi of the right-wing Bharatiya Janata Party as Prime Minister in 2014, the political mood in India has shifted decisively away from socialism and public-sector industry. This has been part of a worldwide trend also reflected by the failure and collapse of the Soviet Union, the opening of China to capitalistic investment, Reaganism in the United States, Thatcherism in Britain, and globalization. In his first Independence Day address as Prime Minister, Modi announced that the Planning Commission, a Nehruvian holdover, would be disbanded. Since 1991, the Planning Commission had continued to make five-year plans, but with little real authority. Following

Biggest Arms Importer, India Wants to Buy Local," *New York Times*, March 6, 2014; Sushant Singh, "SIPRI Data Shows India World's Biggest Arms Importer at Three Times of China," *IE*, March 16, 2015; Stockholm International Peace Research Institute, "International Arms Transfers and Developments in Arms Production," *SIPRI Yearbook 2019*, https://sipri.org/yearbook/2019/05.

[18] Guha, *India After Gandhi*, 683–85.

Modi's announcement, the Twelfth Five-Year Plan (2012–2017) was allowed to run its course, but it was not succeeded by a Thirteenth Plan. India's tryst with centralized economic planning, launched with much fanfare in 1951, ended with a whimper 66 years later.[19]

In place of centralized planning and public-sector industry, Modi promoted foreign private investment through his "Make in India" initiative. This represented a broad shift in the public mood away from the nonalignment period's quest for autarky through domestic production and public-sector investment. Although bastions of socialist ideology persisted in India, including among some Congress Party members and at the famously left-wing Jawaharlal Nehru University in New Delhi, Indian culture, at least among the upwardly mobile middle classes, began to favor the private sector and foreign brands above all. A domestic car brand such as Maruti or Tata was not as desirable as a Skoda or an Audi, even though most foreign-branded cars sold in India were nevertheless manufactured domestically.[20]

A prime destination for the upwardly mobile classes of the early twenty-first century was Gurgaon, located just outside of Delhi National Capital Territory in the state of Haryana. Little more than a village and industrial suburb of Delhi for most of its history, Gurgaon's population exploded after the late 1990s thanks to a revision in Haryana's state tax codes. Within fifteen years, Gurgaon had mushroomed into a city with a population of two million. Gurgaon's urban environment, a riot of office towers with showy rooflines, luxury hotels, high-rise housing developments, capacious air-conditioned shopping malls, and even a small metro railway—not to mention badly-paved, chaotic streets and workers' slums—was built almost entirely by private capital. Unlike Delhi just across the state line, which expands gradually and methodically under the guidance of the governmental Delhi Development Authority, Gurgaon has grown up virtually without a plan. Despite its obvious shortcomings (obvious at least to this author), the upwardly mobile classes speak of

[19] "Modi Makes a Clean Break, Ends Plan Era, Wants to Build Team India," *TOI*, August 16, 2014.

[20] "FDI Means First Develop India, PM Tells Bizmen," *TOI*, September 26, 2014; "'Make in India' Tableau Steals the Show in Republic Day Parade," *TOI*, January 26, 2015, http://timesofindia.indiatimes.com/india/Make-in-India-tableau-steals-the-show-in-Republic-Day-parade/articleshow/46018355.cms (accessed February 3, 2016).

Gurgaon as a great triumph of private capital, achieving rapid urbanization in a way that the public sector's bureaucrats and committees could only dream about.[21]

It has been the central premise of this book that the public sector in India cannot be so easily dismissed, either in the present or in historical analysis. Even to this day, the public sector makes up a significant part of the Indian economy. Beyond the usual roles of policing, transit, highways, education, and other public goods, undertakings of the central and state governments still include the national rail network, heavy manufacturing, steel, telecommunications, and even film production. Although the Nehruvian command economy has long been dead, the public-sector undertakings that it created are still numerous and economically important.

Development during the command-economy years built up the industrial capacity of India and also helped expand technical competence in the country. Technical advisers from eastern and western Europe, North America, and Japan visited India for short or long terms to share their expertise in such fields as nuclear power, aeronautics, dam construction, and heavy manufacturing. Indian citizens also traveled to the donor countries to receive training there. For broader theoretical knowledge, Indians studied domestically at the Indian Institutes of Technology and abroad at places like the Massachusetts Institute of Technology in the United States.[22]

Thanks to this technical competence, the Indian public sector was able to undertake some high-tech endeavors during the command-economy period. The atomic bomb project (discussed in Chapter 8) was one such undertaking. Another was India's space program, which launched its first satellite into orbit in 1980.[23] As this book has described, there were also less-spectacular bridges, dams, and factory projects from this time that

[21] Alex Tabarrok and Shruti Rajagopalan, "Opinion: Designing Private Cities, Open to All," *New York Times*, March 16, 2015, https://www.nytimes.com/2015/03/17/opinion/designing-private-cities-open-to-all.html?/r=0 (accessed June 26, 2020). Gurgaon ("Jaggery-sugar Village") was officially renamed Gurugram ("Village of the Guru") in 2016.

[22] For Indians studying at MIT, see Ross Bassett, *The Technological Indian* (Cambridge, MA: Harvard University Press, 2016).

[23] Gopal Raj, *Reach for the Stars: The Evolution of India's Rocket Programme* (New Delhi: Viking, 2000), 70.

had their own roles to play in the Indian economy and technological landscape.

Autarkical programs and Cold-War politics limited foreign access to Indian markets. Although this blocked some technology from circulating in India, this was not always a bad thing. As Ramachandra Guha notes, IBM's expulsion from India by the Janata government proved to be propitious for India's domestic computing industry. During the eighties, Indian computing companies were able to grow and develop in a protected environment without fear of getting crushed by IBM or other international behemoths. When the Indian market reopened in the nineties, domestic computing companies were big enough by then to be competitive on the world market. The Indian information technology sector has played a major role in the global economy ever since.[24]

In case this sounds like an apology for command-economy socialism and authoritarian high-modernism, I would like to emphasize that this is not my intention here. Nehruvian development had some serious shortcomings and failures, which I have no intention of obscuring. In its attempts to bend nature to accommodate humanity's needs and wishes, Nehruvian modernization caused serious environmental problems: natural habitat loss, destruction of biodiversity, salinization of soil, and air pollution. Even its benefits were unequally distributed. Directed as it was by urban elites, modernization failed to help the poorest and most vulnerable of the Indian population. Indeed, it would be these very people—the rural poor, many of whom were indigenous—who would be disproportionately displaced by the development projects supposedly built for the benefit of all Indians. No one knows exactly how many people were displaced by the development projects of the early independence period, but the number almost certainly stands in the tens of millions.[25]

Yet despite these shortcomings, the Nehruvian era had a virtue that has been conspicuously absent in the post-liberalization era: vision. Nehruvian development was at least theoretically for all Indians. Contemporary

[24] Guha, *India After Gandhi*, 687.

[25] In a lecture that the author attended in New Delhi on June 11, 2012, Ashis Nandy stated that 65 million people had been displaced by development since independence. Development refugees are a problem that the global community has done little to address; see Peter Penz, "Dams, Guns and Refugees," *Alternatives Journal* 30, no. 4 (September/October 2004): 8–12.

neoliberalism, by its very design, is not. The planners of early independent India envisioned the state's role as working for the improvement of the lives of all its citizens through the development of the economy and the creation of public works. In post-liberalization India, the state's role is diminished to promoting business for the benefit of the privileged classes and at the expense of the poor. Nehruvian India failed to eradicate poverty, but economic inequality has only increased since liberalization.

Command-economy socialism seems passé now, as well it should after its serious failures in India and other countries. But its vision, of the state promoting the welfare of its citizens, is as relevant now as it was sixty years ago in the era of nonalignment and five-year plans.

References

1. Bassett, Ross. "Aligning India in the Cold War Era: Indian Technical Elites, the Indian Institute of Technology at Kanpur, and Computing in India and the United States." *Technology and Culture* 50, no. 4 (October 2009): 783–810.
2. Bidwai, Praful. "India's Nuclear Neros." *Outlook*, March 28, 2011. https://www.outlookindia.com/magazine/story/indias-nuclear-neros/270948 (accessed February 24, 2020).
3. Cullather, Nick. *The Hungry World: America's Cold War Battle against Poverty in Asia*. Cambridge, MA: Harvard University Press, 2013.
4. Dhanjal, Gursharan S. *Tarapur: The Politics of Nuclear Age*. New Delhi: Rajdhani Book Service, 1992.
5. Engerman, David C. *The Price of Aid: The Economic Cold War in India*. Cambridge, MA: Harvard University Press, 2018.
6. Guha, Ramachandra. *India After Gandhi: The History of the World's Largest Democracy*. New York: Ecco, 2007.
7. Gupta, Shekhar. *Assam: A Valley Divided*. New Delhi: Vikas Publishing House, 1984.
8. Kudaisya, Ganesh. "Death of a Democrat." *History Today*, May 21–27, 2014.
9. Lal, Sham, ed. *The Times of India Directory and Yearbook Including Who's Who 1969*. Bombay: Times of India Press, 1969.
10. Lal, Sham, ed. *The Times of India Directory and Yearbook Including Who's Who 1973*. Bombay: Times of India Press, 1973.
11. Long, Clarence D. "Nuclear Proliferation: Can Congress Act in Time?" *International Security* 1, no. 4 (Spring 1977): 52–76.
12. Mukherji, Rahul. "India's Aborted Liberalization—1966." *Pacific Affairs* 73, no. 3 (Autumn 2000): 375–92.

13. "Nuclear Non-Proliferation Act of 1978, Public Law 95–242." *The American Journal of International Law* 72, no. 3 (July 1978): 712–42.
14. Penz, Peter. "Dams, Guns and Refugees." *Alternatives Journal* 30, no. 4 (September/October 2004): 8–12.
15. Raj, Gopal. *Reach for the Stars: The Evolution of India's Rocket Programme*. New Delhi: Viking, 2000.
16. Raychaudhuri, Tapan, Irfan Habib, Dharma Kumar, and Meghnad Desai. *The Cambridge Economic History of India*. 2 vols. Cambridge: Cambridge University Press, 1982–83.
17. Seth, Satish C. "'Public Sector and Co-operation': As Seen in the Draft of the Indian Fourth Five-Year Plan (1968–74)." *Annals of Public & Co-operative Economy* 40, no. 4 (October 1969): 487–501.
18. Stockholm International Peace Research Institute. "International Arms Transfers and Developments in Arms Production." *SIPRI Yearbook 2019*. https://sipri.org/yearbook/2019/05.
19. Stockholm International Peace Research Institute. "International arms transfers." *SIPRI Yearbook 2014*. https://sipri.org/yearbook/2014/05.
20. Tomlinson, B.R. "Foreign Private Investment in India 1920–1950." *Modern Asian Studies* 12, no. 4 (1978): 655–77.

Index

0–9
1962 Sino-Indian War, 103–105, 110–113
1965 Indo-Pakistani War, 95, 129, 199
1971 Indo-Pakistani War, 263

A
All-India Trade Union Congress, 245
An-12 transport, 125
Assam, 135
  and Partition, 142
  map, 141, 143
  railways in, 140
Assam Rail Link, 142, 144, 161
Assam State Electricity Board, 169, 177
  transmission lines, 185
atomic research in India, 204–205

B
B-24 Liberator bomber, 82
Bandung Conference, 47
Bechtel Corporation, 223, 225, 226, 249
Belgrade Conference, 49
Bhabha, Homi J., 128, 204
Bhakra-Nangal Project, 1–4, 157
Bharat Heavy Electricals Ltd. (BHEL), 32
Bhargava, Kapil, 97
Bhilai steel plant, 64
Bhopal, 33, 264
Birla family, 19
Boisar, 230
Bokaro Steel Plant, 191
Bombay Labour Union, 223, 246, 249, 250, 252
Bombay Plan, 51
Bowles, Chester, 59, 115, 117, 118, 130
Brahmaputra River, 143
  Bogibeel Bridge, 163
  ferries across, 144, 165
  Jogighopa Bridge, 163

W. A. T. Logan, *A Technological History of Cold-War India, 1947–1969*, Palgrave Studies in the History of Science and Technology,
https://doi.org/10.1007/978-3-030-78767-7

second Saraighat Bridge, 265
Braithwaite, Burn and Jessop (BBJ), 137, 151, 154
Brezhnev, Leonid, 65
Bulganin, Nikolai, 63
Bunker, Ellsworth, 59

C

C-119G transport, 83, 84
C-130 Hercules transport, 124, 125
Canada
and aid for Indian nuclear power, 217
and aid for nuclear power, 217
and financing for Umtru Dam, 175
Canada-India Reactor (CIRUS), 217
Carlson, Roy W. (American engineer), 184
Central Water and Power Commission, 172
Central Water and Power Research Station, 182
Chakravarti, M.N. (nuclear engineer), 206, 207, 216
Chavan, Y.B., 130
Chicago Bridge & Iron, 231
China, 103
border dispute with India, 106–110
first atomic bomb test, 128
map of border with India, 108, 109
Communist Party of India, 50, 64
Congress Party, 223, 263
Czechoslovakia, 44, 169

D

Damodar Valley Corporation, 174
Dayal, Maheshwar (nuclear engineer), 206, 236
de Havilland Vampire fighter, 84
de Pahlen, Serge, 154, 155
Department of Atomic Energy, 226, 256
development refugees, 273

E

Egypt, 95
Eisenhower, Dwight D., 55
visit to India, 56
Emergency, 263
Exercise Shiksha
air operations in, 120–121, 123
American personnel in, 119
American preparations for, 118
planning for, 115
radar sets for, 116, 117, 132

F

F-100 Super Sabre fighter, 117, 122
F-104 Starfighter, 94
Fernandes, George, 223, 246–247, 251–253
Five-Year Plans, 25
First Five-Year Plan, 11, 26, 143
Fourth Five-Year Plan, 5, 11, 40, 262
Second Five-Year Plan, 27, 148
Third Five-Year Plan, 5, 24, 27, 36, 206, 262
Folland Gnat fighter, 83, 84, 87
Orpheus 701 engine for, 90

G

Galbraith, J.K., 59, 60, 115, 220
Gandhi, Indira, 60, 63, 255, 262–264
Gandhi, Mohandas K., 15, 20, 245
Gandhi, Rajiv, 270
Ganguli, B.C. (railway engineer), 137, 148, 152, 156, 158
General Electric, 225, 230, 239

bidding for Tarapur contract, 210–213
selection as prime contractor for Tarapur project, 213, 215
Graham, Gordon M. (USAF Brigadier General), 118, 122, 124
Green Revolution, 25, 264
Gurgaon (Gurugram), 271–272

H
HA-300 fighter, 96–98
Heavy Electricals (India) Ltd. (HE(I)L), 32, 38
heavy electricals industry, India, 32–36, 266
HF-24 Marut fighter, 268
and Indian industry, 99
development of, 85–89
engine for, 89–91
photo of, 87
proposed technical collaboration with United States, 93
technical collaboration with Egypt, 95–98
Hind Mazdoor Sabha, 246
Hindustan Aircraft Ltd. (HAL), 80–82, 84
and HF-24 project, 85, 89, 93
Gas Turbine Division, 90
manufacture of MiG-21 fighters, 91
Hindustan Construction Co. (HCC), 137, 149, 150, 158
Hindustan Motors, 32, 164
Hippie Trail, 68
Hirachand, Seth Walchand, 80, 149

I
IBM, 273
import substitution, 17, 31, 263, 270
Inagua Gull, 231
India
and American culture, 66
coercive power of government, 247–249
dams in, 172–175
electrification, 173
embassy in Washington, 51
popular perceptions of the United States, 124
population growth, 262
treaty with Soviet Union, 70
Indian Air Force
after Sino-Indian War, 113
and air control, 78
and American technology, 80
and British technology, 80
and importing fighters, 78
and Russian technology, 92
founding, 78
in World War II, 80
Indian Journal of Power and River Valley Development, 174
Indian National Trade Union Congress, 223, 246, 249, 252
Indian Railways, 138–142
India Supply Mission, 51
INS Vikrant, 39
International Atomic Energy Agency (IAEA), 218, 219, 267

J
Jaduguda Mines, 205, 215
Janata Party, 31, 263, 273
Japan, 189
Jogighopa, Assam, 147
Johnson, J.W. (American engineer), 228

K
Kapoor, Raj, 67
Kashmir, 74
Kennedy, Jacqueline, 60

Kennedy, John F., 52, 58–61, 104, 127
Khadi, 24
Khasi Hills, 169, 177, 185
Khasi people, 176–177
Khrushchev, Nikita, 63
Kirloskar, S.L., 51
Kosygin, Alexei, 65
Krishna Menon, V.K., 50, 90, 104
Kuljian Corporation, 227–228

L
labor in India, 245–247
LCA (Light Combat Aircraft) Tejas, 269

M
Mahalanobis, P.C., 4, 27
Maharashtra State Electricity Board, 230
Malaysia, 36
Manhattan Project, 203
Meghalaya Plateau, 147
Messerschmitt, Willy, 96
MiG-21 fighter, 77, 91, 121
Modi, Narendra, 270–271
Mokameh Ghat Bridge, 149, 151

N
Nagarjuna Sagar Dam, 37
Nasser, Gamal Abdel, 30, 47, 95
National Planning Committee, 25
Nehru, B.K., 52
Nehru, Jawaharlal, 22, 60
  and 1962 Sino-Indian War, 104, 111
  and Brahmaputra Bridge project, 149
  and defense production, 76
  as Foreign Minister, 44
  chairman of National Planning Committee, 25
  death, 261
  dedication of Bhakra Dam, 2
  first visit to the United States, 52
  international travels, 43
  views on science and technology, 23
  visits to the Soviet Union, 61, 63
  visits to the United States, 52
Nichols-Roy, Stanley, 178
Nixon, Richard, 263
Northeast Frontier Railway, 137, 145
nuclear fuel in India, 205–206, 215
nuclear power in India, 258
nuclear reactors
  Apsara research reactor, 204
  Canada-India Reactor (CIRUS), 205, 267
  reactor types, 211, 226
nuclear weapons in India, 202–203

P
Pakistan, 46
  and American military aid, 83
Pakistan Air Force, 94
Palam airfield, 117
Pandit, Vijaya Lakshmi, 62
Partition, 73–75, 142
Passman, Otto E., 59
Peaceful Nuclear Explosion (1974), 266
Pinto, E.W. (IAF Air Vice Marshal), 118, 127
PL-480 (Food for Peace), 56
Planning Commission, 2, 26, 262
policing in India, 223, 248–250
Portlock, R.L. (Bechtel manager), 252
Premier Automobiles, 32

R
Rajasthan Atomic Power Station, 35, 217, 226
Ram, Jagjivan (Railway Minister), 145, 148

S
Saraighat Bridge, 136, 265
  construction, 157–162
  dedication by Nehru, 162
  diagram, 159
  opening, 161
  photo, 162
  question of siting, 146–148
  seismic design, 159
  structural design and analysis by Freeman Fox and Partners, 157
  tender process for second contractor, 151–157
Sharavathi Hydroelectric Project, 175
Shastri, Lal Bahadur, 261
Siliguri Corridor, 135
Singh, Arjan (IAF Air Vice Marshal), 114, 115
Singh, Karnail (Railway Board), 137, 142, 148, 154
Six-Day War, 98
Slocum, Harvey (American engineer), 2
Soviet Union/Russia
  as arms dealer for India, 269
  development aid to India, 64
  industrialization, 28
  relations with India, 61–66
Srinivasan, M.R. (nuclear engineer), 206, 207
Stalin, Joseph, 62
Su-7 fighter, 98
swadeshi, 24
Swatantra Party, 65, 252

T
Tank, Kurt, 85–87
Tarapur Atomic Power Station
  1965-66 labor strike, 223–225, 249–253
  and children, 243–245
  and gender, 242–243
  and industrial paternalism, 240–241
  as symbol of modernity, 257
  beginning of construction, 221
  bidding for contract, 210–214
  construction, 230–233, 253–254
  fuel supply, 215–219, 266–267
  housing colonies, 238–240
  in 1965 war with Pakistan, 199–200
  operating problems, 254–255
  parts from the United States, 230
  reasons for construction, 206–207
  safety concerns, 256, 267–268
  site selection, 207
  training for operators, 235–236
  workforce, 233–235
Tarapur Reporter, 237
Tarapur Swim and Racquet Club, 241–242
Tashkent Conference, 65
Tata Iron and Steel Company, 19, 37
Tata, Jamsetji, 19
Taylor, Maxwell, 127
technocracy, 28
test pilots, 97
textile industry, 18
Tezpur, 104, 145
Tito, Josip Broz, 48
Truman, Harry S, 55

U
Umiam Hydroelectric Project, 265
  commissioning, 193
  construction, 181–186
  diagram, 183

financing by United States government, 169, 187–192
land acquisition for, 175–179
map, 180
pollution in reservoir, 195
Umtru Hydroelectric Project, 175
Union Carbide, 264
United States
and nuclear fuel safeguards, 218–219
and romanticization of India, 67–69
embassy in New Delhi, 56
Military Assistance Program to India, 111, 114, 124, 130–132
popular perceptions of India, 52, 123, 125
relations with India, 45, 50–61, 69
supplier of construction machinery for Umiam Dam, 188
United States Agency for International Development (USAID), 59, 187, 191–192, 214, 227, 230, 263
United States Air Force, 111, 115, 121, 124
urbanization, 16

W
Warren truss, 160, 161
World War II, 53, 140

Y
Yugoslavia, 48, 170

Z
Zhou Enlai, 48, 111